utledge Environmental Humanities
ries editors: Scott Slovic (University of Idaho, USA),
ni Adamson (Arizona State University, USA) and
ki Masami (Aoyama Gakuin University, Japan)

Routledge Environmental Humanities series is an original and inspiring ven- recognising that today's world agricultural and water crises, ocean pollution resource depletion, global warming from greenhouse gases, urban sprawl, population, food insecurity and environmental justice are all *crises of culture*.

Ecological Ambivalence, Compl
and Change

This book provides a systematic, interdisciplinary analysis o
and tensions associated with today's ecological transform
an Environmental Humanities perspective. It explores the
ambivalence, where conflicting reactions, beliefs, or feeling
icies or private practices for "saving planet Earth" threaten to

Under the umbrella of the Environmental Humanities, the
scholars from fields such as environmental history, ecologic
geography, and ecocriticism. Contributions investigate
ambivalences, wound up with processes of environmental
conceptually and empirically. Case studies range from wi
green mineral mines in Mexico, and from chemical contami
Rocky Mountain Arsenal in Denver, USA. Additionally, wit
environmental communication—as in Philippe Squarzoni's
Changed or G'Ebinyo Ogbowei's poetry—contributions
pathways for overcoming ambivalences, managing them cr
the concept as whole. The volume highlights how the huma
social sciences can work together to help humankind dev
skills to overcome paralysis and engage in practical action,
forth ambivalence as an approach for being in today's world

This book will be of interest to researchers, academics,
Environmental Humanities, the social sciences, the humani
mental sciences. It will also be useful for decision-makers, t
activists.

Simone M. Müller is DFG Heisenberg Professor of Global
and Environmental Humanities at the University of Augsbu

Matthias Schmidt is Professor of Human Geography and Tr
at the University of Augsburg, Germany.

Kirsten Twelbeck is an American Studies scholar and coord
doctoral program Re-Thinking Environment, a cooperation
of Augsburg and Ludwig Maximilian University of Munich

The reality of understanding and finding adaptive solutions to our present and future environmental challenges has shifted the epicenter of environmental studies away from an exclusively scientific and technological framework to one that depends on the human-focused disciplines and ideas of the humanities and allied social sciences.

We thus welcome book proposals from all humanities and social sciences disciplines for an inclusive and interdisciplinary series. We favour manuscripts aimed at an international readership and written in a lively and accessible style. The readership comprises scholars and students from the humanities and social sciences and thoughtful readers concerned about the human dimensions of environmental change.

For more information about this series, please visit: www.routledge.com/Routledge-Environmental-Humanities/book-series/REH

Ecological Ambivalence, Complexity, and Change

Perspectives from the Environmental Humanities

Edited by Simone M. Müller, Matthias Schmidt and Kirsten Twelbeck

Routledge
Taylor & Francis Group
LONDON AND NEW YORK

earthscan
from Routledge

First published 2025
by Routledge
4 Park Square, Milton Park, Abingdon, Oxon OX14 4RN

and by Routledge
605 Third Avenue, New York, NY 10158

Routledge is an imprint of the Taylor & Francis Group, an informa business

British Library Cataloguing-in-Publication Data
A catalogue record for this book is available from the British Library

ISBN: 978-1-032-62794-6 (hbk)
ISBN: 978-1-032-62797-7 (pbk)
ISBN: 978-1-032-62798-4 (ebk)

DOI: 10.4324/9781032627984

Typeset in Times New Roman
by Newgen Publishing UK

Contents

Figures

Contributors

Maria Backhouse is Professor of Environmental Sociology with a focus on socio-ecological transformation, resilience design, and climate at the University of Augsburg. Her research interests encompass the political ecology of climate change, global socio-ecological inequalities, and conflicts surrounding socio-ecological transformation.

Angeliki Balayannis is Associate Professor in the Knowledge, Technology, and Innovation Group at Wageningen University & Research, the Netherlands. She is a transdisciplinary social scientist specializing in the problems of pollution, waste, and planetary health. Her work is conceptually located at the intersections of geography and feminist science and technology studies, which enables her to foreground questions on epistemic in/justice. She is currently co-producing epistemic interventions at the science–policy interface, contributing to the establishment of a new intergovernmental panel on chemicals, waste, and pollution prevention (a panel whose name is under negotiation at the time of publication).

Helena Bilandzic is Professor in the Department of Media, Knowledge, and Communication at the University of Augsburg, Germany. Her research interests include media effects related to science, the environment, and health, as well as social and moral issues, narrative experience, and persuasion. Her current research projects investigate the "Effects of narrative evidence in science news coverage of genomic research" (funded by the German Research Foundation), analyze "De- and restabilisation of evidence in the COVID-19 crisis" (funded by the German Research Foundation), and explore "Science communication about and with communicative artificial intelligence: emotions, engagement, effects" (funded by the German Federal Ministry of Education and Research). Helena was an elected member of the review board for Social Sciences of the German Research Foundation and serves on the editorial boards of several scholarly journals.

Iris Borowy is Distinguished Professor at the University of Shanghai and Director of the Center for the History of Global Development. She has published widely on topics of global health, sustainable development, and international organizations. She is the author of *Defining Sustainable Development for Our*

Common Future. A History of the World Commission on Environment and Development (Brundtland Commission) (Routledge, 2014) and co-editor of the *Routledge Handbook on the History of Development* (2022). She is founding co-editor of the serial *Yearbook of the History of Global Development* (de Gruyter).

Sule Emmanuel Egya is Professor of African Literature and Environmental Humanities at Ibrahim Badamasi Babangida University, Lapai, Nigeria. His research interests include literature and the environment, African migration writing, knowledge production in Africa, and decolonial discourse. His current research examines environmental imagination in African 20th-century literature. He is the author of many articles that have appeared in journals in Africa, Asia, Europe, and North America. His books include *Nation, Power and Dissidence in Third Generation Nigerian Poetry in English* (NISC, 2014, 2019); *Niyi Osundare: A Literary Biography* (SevHage, 2017); *Power and Resistance: Literature, Regime, and the National Imaginary* (SevHage, 2019); and *Nature, Environment and Activism in Nigerian Literature* (Routledge, 2020). He has co-edited *Studies in Scientific and Cultural Ecology* (SevHage, 2021) and *Orality, Textuality, Society: New Perspectives on Nigerian Literature and Culture* (SevHage, 2023). He also writes fiction and poetry under the pen-name "E. E. Sule." He is the author of the novels *Sterile Sky* (winner of the Commonwealth Book Prize Africa Region, 2013) and *Makwala* (ANA Prose Prize, 2019), as well as the poetry collection *What the Sea Told Me* (winner of the ANA Gabriel Okara Prize, 2009).

Victor A. Ferretti is Professor of Iberian and Latin American Studies at the University of Augsburg, Germany. His research focuses on early-modern discursivity, urban cultural studies, and environmental aesthetics. He is co-editor of the *Studia Augustana* (De Gruyter) publication series, with medievalist Klaus Wolf, and author of a monograph on the media cultural history of bucolic poetry, *Bucolica mente* (Studia Romanica Chiloniensia, 2017), as well as articles on fiction theory, literary phenomenology, poietic world-making, etc.

Rafael Hernández Westpfahl is a PhD candidate at the Heidelberg Center for Ibero-American Studies, University of Heidelberg, with expertise in the political ecology of natural resources. His research focuses on the governance and spatial dynamics of mining in Mexico. His dissertation project is funded by the Landesgraduiertenförderung Baden-Württemberg, and his fieldwork in Mexico was funded by the German Academic Exchange Service (DAAD).

Rocío Hiraldo is a post-doctoral researcher in the Social Anthropology Department at the University of Sevilla. Her research concentrates on the intersections between labor, class relations, and environmental protection through a focus on working-class people's experiences of and responses to ecological and climate transitions. Her work has been published in *Environmental Planning E*, *Review of African Political Economy*, *Journal of Agrarian Change*, and *Human Geography*.

Anja Kalch is Senior Researcher and Program Coordinator in the Department of Media, Knowledge, and Communication at the University of Augsburg, Germany. Her research interests include health communication, environmental communication, narrative persuasion research, and media reception and effects. She is co-principal investigator of several research projects, including "Development and evaluation of holistic, indicator-based sustainability instruments in acute inpatient health care" (funded by the Bavarian State Ministry of Health and Care) and "Evidence-based health information for patients with pulmonary embolism in the post-acute treatment phase" (funded by Federal Joint Committee for the Promotion of Health Services Research). She serves as co-chair of the section "Media reception and effects" of the German Communication Association.

Sebastian Lundsteen holds a PhD in Environmental History from The Greenhouse—Center for Environmental Humanities at the University of Stavanger. His dissertation explores the intersection between chemical pollution, critical theory, and environmental justice in the context of the welfare state in the Nordic region.

Simone M. Müller is DFG Heisenberg Professor of Global Environmental History and Environmental Humanities at the University of Augsburg. Bringing together technology, the economy, and the environment, she works at the intersection of environments above and below ground, wetlands from the quaking zone, and legacy sites as toxic commons. Her research has received numerous awards and fellowships, among them from the Smithsonian Institution, the Science History Institute, and the University of Pennsylvania. She is editor of the series *Ecology and History* at Ohio University Press.

Ulrike Ohl is Professor and Head of the Chair of Geography Education at the University of Augsburg, Germany. Her research focuses on professionalizing future geography teachers, education for sustainable development, transformative learning, and climate change education. Research projects under her supervision aim to connect theoretical and practical perspectives; intervention research and design-based research are central approaches in this regard.

Nisha Paliwal is a research scholar in the School of Liberal Arts at the Indian Institute of Technology, Jodhpur, India. She is interested in folklore and community practices of the Thar Desert. Her PhD research area stands at the intersection of energy humanities and political ecology, where she studies the implications of green energy transition/intervention at sacred sites called "Orans" in the Thar desert of Rajasthan. Her research work involves understanding the effects of such transitions and interventions on local practices as well as desert ecologies. She is also developing a digital museum for the Orans of Rajasthan as a means to document local ecological knowledge systems and traditional cultures.

Dibyadyuti Roy has been a researcher and educator in multidisciplinary academic environments across three different continents. His scholarship and teaching interrogates cultural narratives of/about dominant technologies, their genealogies, mediations, and fallouts. Currently, he is Assistant Professor in Cultural Studies, Media Studies, and Digital Humanities at the University of Leeds, where he is also Program Director of the BA in Cultural and Media Studies. Dibyadyuti Roy is a research affiliate of the Data and Society Research Institute (USA) and a founding member of India's first Digital Humanities collective: the Digital Humanities Alliance for Research and Teaching Innovations (DHARTI).

Kerstin Schlögl-Flierl is Professor of Moral Theology in the Catholic Theological Faculty at the University of Augsburg. Her research is on artificial intelligence and environment and climate issues. She is head of the interdisciplinary master's program Environmental Ethics. Since 2020, she has been a member of the German Ethics Council and was Chair of the Working Group on Climate Justice. Her research interests are on environmental medicine ethics, climate ethics, and water justice.

Matthias Schmidt is Professor of Human Geography and Transformation Research at the University of Augsburg. His research centers on human–environment relations, political ecology, resource geography, and mountain studies, with a regional focus on the Global South, in particular Central Asia (Kyrgyzstan) and South Asia (Pakistan). He is speaker of the international doctoral program Re-Thinking Environment. The Environmental Humanities and the Ecological Transformation of Society and for and board spokesperson for the Environmental Science Center at Augsburg University. Together with Hubert Zapf, he edited *Environmental Humanities: Beiträge zur geistes- und sozialwissenschaftlichen Umweltforschung* (V&R Unipress, 2021).

Dietmar Süß is Professor of Modern and Contemporary History at Augsburg University. His research focuses on the history of the labor movement in the 19th and 20th centuries, the history of violence, and national socialism. His most recent publication is *Solidarity. From the 19th Century to the Corona Crisis* (with Cornelius Torp, Bonn 2022).

Kirsten Twelbeck is Coordinator and Advisor for the international doctoral program Re-Thinking Environment, a cooperation between the University of Augsburg and LMU Munich. She has a PhD and a "habilitation" in American Studies, focusing on modes of literary representation in contexts of political and social transformation. After several visiting professorships, she joined the Environmental Science Center at Augsburg University to help establish the Environmental Humanities as an interdisciplinary research field. She has taught and published widely in the field of Environmental Humanities and currently works on a monograph centering on wheat.

Marcus Wagner is Professor of Management, Innovation and Sustainable Business at the University of Augsburg and Associate Member of the Bureau d'Economie Théorique et Appliquée (University of Strasbourg). His research focuses on innovation, sustainability, and management, and he has been published, inter alia, in *Journal of International Business Studies*, *Journal of Economic Behaviour and Organization*, *Journal of Business Venturing*, and *Research Policy*. He is a member of the board of KUMAS (a transfer organization for sustainability-related knowledge), a principal investigator of his university's international doctoral program Rethinking Environment and founding member of its interdisciplinary Center for Climate Resilience.

Hubert Zapf is an Americanist and Co-Director of Environmental Humanities at the University of Augsburg. He is a founding member of the joint Augsburg-Munich international doctoral program Rethinking Environment: The Environmental Humanities and the Ecological Transformation of Society. He has published widely on cultural ecology, English and American literature, and literary history and theory. His recent publications include *Literature as Cultural Ecology: Sustainable Texts* (Bloomsbury, 2016); *Handbook of Ecocriticism and Cultural Ecology* (ed.) (De Gruyter, 2016); Ecology in American Literature (with Timo Müller), *Oxford Research Encyclopedia of Environmental Science* (Oxford University Press, 2022); "Posthumanism or Ecohumanism: Environmental Studies in the Anthropocene," *Journal of Ecohumanism* 1 (1) 2022: 5–17.

Acknowledgments

We would like to thank the Elite Network of Bavaria (ENB) for their generous support of this book project.

Introduction

Simone M. Müller, Matthias Schmidt, and Kirsten Twelbeck

In 2021, the Central American nation of Belize faced state bankruptcy. Yet, just before foreclosure, the country received a loan from the US-based NGO The Nature Conservancy in a debt-for-nature swap scheme. Through this transaction, Belize not only recapitalized its national budget, thus lowering its foreign debt, but it also committed itself to spend about US$ 4 million a year on marine conservation until 2041, double its marine-protection parks, and continue conservation after 2040 through an endowment fund of US$ 23.5 million.[1] Worldwide, the scheme triggered new discussions and renewed hope on how the coupling of fiscal policy and nature conservation could be a way forward to respond to two great issues facing the majority of the world: environmental degradation and debt.[2] At the same time, the deal received no univocal praise. Working through the ambivalences of the scheme, Jonas Gamso, writing for the *Standford Social Innovation Review*, still posed the following question: "Debt-for-Climate Swaps Can Save the Planet. Why Aren't They?"[3]

Debt-for-nature swaps are the exchange of a debtor country's external obligations for the country's agreement to use that money to support conservationist projects.[4] They exist in two formats, either as a three-party swap, where an NGO buys the debt from a creditor country and then re-sells it at a discount to the debtor country, or as a two-country swap between a debtor and a creditor country. In a bilateral swap, the creditor country decides the policy agenda without extensive input from the debtor. This is also the source of the scheme's greatest critique. Heralded as a win-win for proponents, opponents target the scheme as a sell-out of national sovereignty with actors from the West telling "the rest" how to protect nature. Experiences have revealed the deals as being slow, cloaked in secrecy, and administratively inherently complicated. In the end, debt-for-nature swaps remain an ambivalent tool whereby both public and political assessments swing back and forth: is this a way to achieve both the environmental and fiscal stability of developing countries, or is it just another form of ecological imperialism?[5]

Set in motion by climate change, biodiversity loss, or planetary pollution, the call for socio-political and ecological transformations has become increasingly audible in political and public discourse, culminating in the persistent demand for

DOI: 10.4324/9781032627984-1

far-reaching measures to secure the survival of future generations. Initially, a wide variety of actors seem to agree on the necessity of radical socio-ecological changes to combat fatal environmental issues. Yet, this consensus crumbles once concrete measures are put on the table—such as debt-for-nature swaps as well as alternative schemes relating to energy, resource use, or waste disposal—as any change may entail unforeseen difficulties or create new contradictions, even among a coalition of the willing. In some cases, national implementation fails international consensus. In other cases, local urgency is obscured by the planetary scale or temporal dimension of environmental problems.[6] The great complexity of today's socio-environmental crises makes it difficult to take actions that are supported by as many people as possible. Often, serious tensions arise over costs, responsibilities, and accountabilities, also because their distribution among different stakeholders hinges on centuries of uneven growth and extraction. The ecological transformation is also a highly political endeavor, where not only different worldviews, but also different worlds seem to clash.[7] This, in turn, adds to widespread frustration and potentially reinforcing rhetorical defense tactics, narratives, and attitudes that also help modern societies circumvent the socio-environmental dilemmas that mark many of their actions. Ultimately, and particularly in line with the accusations of NGOs such as Fridays for Future, Extinction Rebellion, or the Last Generation, humanity seems to achieve too little too late despite trying.[8]

Instead of despairing about this seemingly locked-in situation, one may look at it as typically human. Both individuals and groups perceive, experience, and react to a situation in surprisingly different ways. Social and cultural factors, individual experience and family relations, neurological processes, and even genetics[9] contribute to the construction of a reality that is, with regard to societies at large, multiple and selective and therefore intrinsically ambivalent. While awareness of the multivocality and inner, societal, and even global conflicts that go along with ecological crisis has made its way into different social science and humanities disciplines, the notion of ecological ambivalence has not yet been addressed systematically from an interdisciplinary viewpoint. Its possibly creative potential for imagining alternative futures has remained underexplored. This edited volume, *Ecological Ambivalence, Complexity, and Change: Perspectives from the Environmental Humanities*, seeks to fill this gap and inspire conversations, new approaches, and narratives across disciplines that capture the dilemma in which conflicting reactions, beliefs, or feelings toward public policies or private practices for "saving planet Earth" threaten to produce a stalemate. From the entry point of the critical and interdisciplinary Environmental Humanities paradigm, this book brings together scholars from a large array of research fields and institutions from the humanities and social sciences that are diverse in terms of archives, theories, and methods. Yet, those are joint together in their interest to not only tackle ecological ambivalences through interdisciplinary conversation and collaboration, but also propose ambivalence as an approach to being in today's world. In 13 contributions and concluding remarks that also survey the individual contributions, authors from fields ranging from Environmental History to Ecological Economics, and from Human Geography to Ecocriticism, investigate the dissonances and dilemmas—or, as we say, *ambivalences*—wound up with environmental transformation processes, both conceptually (Part I) and empirically (Part II).

In the first part of the book, the authors work through ambivalences of socio-ecological transformation from their respective disciplinary vantage points and so provide the entry to the larger, more conceptual conversation between different disciplines of the book as a whole. The empirical studies in the second part of the book, in turn, tease out different constellations that produce and sustain dissonant modes of engaging with the planetary crisis and so study ecological ambivalences at work. Case studies range from wind farms in India through green mineral mines in Mexico, from chemical contamination in Denmark through Rocky Mountain Arsenal in Denver, USA, and from labor and environmental activists in Spain and Portugal through bioplastic in German schoolbooks. In so crisscrossing the globe, *Ecological Ambivalence, Complexity, and Change* not only provides a global perspective, but also teases out how ecological ambivalences are often situated between power dynamics and inequalities of a Global North-Global South constellation. Additionally, and with a focus on creative environmental communication and narratives—as in Philippe Squarzoni's graphic novel *Climate Changed* or G'Ebinyo Ogbowei's poetry—the contributions also extrapolate the particular contributions of literature and the arts to live with—and possibly overcome—a stagnant status quo and for imagining alternative futures. Not without critically engaging with the concept of ambivalence as such, authors are interested in how creating and experiencing art and literature, and thinking about these forms of human expression, can contribute to a better understanding of today's environmental challenges. In the end, *Ecological Ambivalence, Complexity, and Change: Perspectives from the Environmental Humanities* is as much an exercise in studying and living with dissonances as it is a field guide to engaging with them conceptually and creatively.

Ambivalence—More Than an Expression of Disorder and Difference

Ambivalence, i.e., the possibility of assigning an object or an event to more than one category, is foremost an expression of unresolved tension. Such tension is a feature of human communication because despite its classifying function, language fails to "achieve some sort of commodious filing cabinet that contains all the files that contain all the items that the world contains—but confines each file and each item within a separate place of its own."[10] In principle, such tensions accompany human action and being, and they are the result of conflicting desires, feelings, and thoughts existing side by side.

While ambivalence is a human reaction to contradictions, and complexity has always existed, the term itself did not emerge until the Swiss psychiatrist Eugen Bleuler established it in 1910 to describe behavioral patterns in schizophrenia.[11] A composite of the Latin terms *ambo* (both) and *valere* (worth, strong), ambivalence has since served not only to explain the collective psychology leading to ruptures in history and culture, but also as an approach and method to re-think potentially problematic, or outdated, societal norms and values. It is hardly surprising, then, that ambivalence has been a key element in the epistemic, ethical, and aesthetic response of the humanities and social sciences to techno-economic modernity. While the natural and technological sciences, in line with a modernist mindset dominating the public discourse, promise to offer unequivocal concepts and unambiguous answers, the humanities and social sciences have cultivated alternative forms of thinking in relational, non-binary, and thus ambivalent ways that question monopolistic claims to truth and certainty via an awareness of their implications in pluriversal connectivities and complex power relations. However, this rich legacy of critical-reflective thinking in ambivalence has so far not been actively employed, let alone systematically addressed. It has only recently begun to find attention in view of the problems emerging from translating ecological ideas into transformative practice.[12]

Nicholas Gamso's question as to why debt-for-climate swaps are not saving the planet illustrates how, in recent years, a particular kind of ambivalence, *ecological ambivalence*, has come to the fore in public and political discussions. The term refers to the disorder, ecological grief, global dissonance, and denial we—both as individuals and as members of society—feel, witness, seek to understand, or combat in the face of multiple and seemingly unresolvable planetary crises. As contributions in this volume illustrate, *ecological ambivalence* works on multiple scales and in multiple formats. It takes the form of the symbolic tree we plant when using transportation high on carbon emissions or of the organic grapes we purchase, delivered from far-flung, water-scarce regions. *Ecological ambivalence* travels with scholars engaged in environmental issues who still fly from all over the world to far-away in-person meetings. More generally, it is the state of inaction and paralysis in the face of environmental damage despite easily accessible knowledge and clear recommendations from the scientific community, alongside the string of new contradictions and difficulties that well-intended actions can easily cause given the complexities of both ecological and socio-political systems. On a systemic level, *ecological ambivalence* is the unresolved tensions attached to setting-up wind farms on indigenous territory, thereby destroying sacred sites and disrupting religious practices or the remediation of contaminated sites without thinking-through the absence of an "ultimate sink."[13] It is the engagement of the World Economic Forum in Davos with matters of climate change, or the involvement of oil-producing countries at COP28, hosted in Dubai, that have observers swaying between greenwashing and the green economy. Finally, it also is the necessary transition to alternative means of mobility, such as e-fuels, that represents a "solution" that opens up just another can of worms.

As we fail to make sense of such complex situations or to choose between alternatives that each represents a catch-22, ambivalence is often accompanied by anxiety and discomfort. In recent years, climate anxiety—*Klimaangst*—has become a studied mass phenomenon, particularly among the younger generations, as well as its counter-form, climate denialism.[14] The state of angst, inaction, and paralysis that affects societies while they should instead be taking action has created a conundrum that the social sciences have set out to solve in their studies on human behavior. Environmental psychologists, neuro-scientists, economists, scholars of communication and, to a degree, philosophers, together with sociologists,[15] are at the forefront of unpacking why individuals and collectives do not attempt to mitigate climate change or even resist necessary transformations: after all, information about the facts is easily accessible, recommendations from the climate sciences are clear, there is a moral imperative based on knowing what is required, and having the freedom to change how we do things is a basic promise in democratic societies.[16] As this scholarship suggests, however, many people do not want to follow what appears to be evidence-based practice. Experimental studies show that those denying climate action often do so to protect their privileges or to avoid uncomfortable emotions, rather than for reasons of ignorance.[17] Additionally, scholars suggest, the human mind rather responds to the challenges of climate change with new narratives of evasion that help sustain old privileges, rather than with behavioral change. This deviation in narratology, in turn, some environmental scholars and activists seek to counter with yet other, perhaps more radically hopeful, narratives, ultimately creating a ping-pong between seemingly incompatible positions.[18]

The notion that transformation measures pose a major challenge to societies has become particularly obvious since the late 19th century, when technological innovation, secularization, and societal changes challenged cultural norms and the way of life in industrialized countries. The enthusiasm for the new and untried went along with deep-rooted anxieties concerning the future. In his elaborations on modernity, sociologist Zygmunt Bauman attests to European societies and politics yearning to diminish the ambivalences of modernization. Technological and cultural transformation processes, along with political instability, fueled the desire for clarity and order, ultimately defining the fight against ambivalence as a characteristic of that era. At the same time, art and literature, as well as other forms of cultural creativity, in their response to the challenges of modernity have emphasized the need to acknowledge ambivalence and ambiguity, not only as an irritable disturbance of logocentric control, but also as an intrinsic part of all human experience, thought, and action.[19] In the field of philosophy, the critique of modernity advanced in the theory of the "dialectic of enlightenment," as advocated by Theodor Adorno and Max Horkheimer, exposed the Janus-faced nature of instrumental reason in its alienating effects on humans and their relationship with the natural world.[20]

Adjusting this dialectic to the era of the Anthropocene, the era when humanity's impact on the planet exceeds that of the sun, Ulrich Beck's concept of reflexive modernity thematized the severe repercussions of accelerated modernization. Beck's world risk society is faced with historically unprecedented risks accompanying its techno-economic advances, such as the threat of nuclear destruction, pollution, or unmanageable global waste.[21] As of the 1960s, and with increasing fervor, influential post-structural theorists such as Jacques Derrida, Michel Foucault, and Judith Butler exposed difference and ambiguity (which refers to something that can be understood in two or more ways or is open to interpretation, like Derrida's neologism "différance") as new ways of reading texts, history, or gender as fluid rather than as fixed entities. The destabilization of established binaries was equally the concern of postcolonial critic Homi K. Bhabha, whose notion of colonial mimicry undid the hierarchization between colonizer and colonized and established the "third space" concept as a site of creative tensions and the negotiation of cultural hybridities.[22] Since the 2000s, an ongoing ontological turn has emphasized how both unity and diversity must be considered simultaneously if we are to understand how not only different worldviews, but also different worlds come into being.[23]

Ecological ambivalence is thus more than an expression of disorder and difference, or merely a critique of binary conceptual thinking. In an era of climate change, biodiversity loss, and planetary contamination, it is an intrinsic part of all human experience, thought, and action—an expression of the current Zeitgeist and logic. It works structurally on the individual as well as the societal level and ranges from cognitive dissonance to the seemingly incommensurateness of even institutions, ideologies, or worlds. To study the complexity of these variants of ambivalences that emerge whenever humans try to define and implement concrete measures to improve the environmental status quo, this volume, *Ecological Ambivalence, Complexity, and Change: Perspectives from the Environmental Humanities*, draws on a rich legacy of critical-reflective thinking in ambivalence. As its two-part structure illustrates, however, it does so, as a confluence of simultaneous developments in different disciplines, employing an interdisciplinary lens that brings contributions from many different disciplines into conversation.

Ambivalence as an Approach of Lived Interdisciplinarity

Many disciplines in the humanities and social sciences rely on sociological and philosophical conceptualizations of ambivalence, ranging from Adorno to Beck, to expose the problematic dualisms that reside in much contemporary thinking and action. With their focus on and consideration of different social groups, political perspectives, economic interests, and spatial dimensions, for instance, the social sciences, including Human Geography, inevitably encounter ambivalences. Individual interests and needs conflict with social necessities, and what is considered necessary at a global level can cause considerable damage at the local. Protecting the rainforest to safeguard biodiversity and mitigate climate change for the benefit of humanity often leads to violence and the displacement of local communities. The call for decarbonization, the use of remote areas as carbon sinks, or safeguarding the global gene pool are loudly articulated by the affluent societies of the Global North, but they are seen by the majority of the world as a form of encroachment on their sovereignty, or even as neo-colonialism.[24] Yet, so far, the scholarly discussion rarely explicitly focuses on ambivalence as a theoretical term that can open up new ways of thinking and instead attempts to explore conflicts that emerge in environmental transformation contexts. *Ecological Ambivalence, Complexity, and Change: Perspectives from the Environmental Humanities* seeks to propose ambivalence not only as a descriptive mode or a tool of analysis, but also as a perspective and an approach of lived interdisciplinarity and of engaging with the world.

Caspar Battegay's remark about the difference between ambiguity and ambivalence sets the tone for this lived interdisciplinarity and engagement: "[…] the observation that our world can be confusingly ambiguous and exhausting is banal. The fact that it is ambivalent, on the other hand, seems somehow interesting."[25]

Battegay is hardly the first to celebrate the invigorating potential of ambivalence. A century before the psychiatrist Eugen Bleuler coined the term ambivalence, poet John Keats admired William Shakespeare's "negative capability" and cast it as a creative method of "being in uncertainties [...] without any irritable searching for fact and reason."[26] In literature, linguistics, and literary studies, the creative potential of ambivalence is actually commonplace—poetry in particular, but also straightforward forms of prose rely on the multilayered nature and ambiguity of language to defy the familiar and convey the idea that there is something other than what we have learned to accept as a given. Ambivalence as an approach, we argue in this book, puts this creative potential to use by highlighting contradictions without smoothing them out, by putting the dilemmas out there as something equally invigorating and disturbing. Only by fully recognizing ambivalence can it become a useful tool in the search for possible pathways that lead the way out of an impossible situation.

The Environmental Humanities as a critical and interdisciplinary field, furthermore, provide a particularly suitable framework to undertake such an intellectual endeavor. With its many different scholarly and institutional points of departure, accompanied by journals, conference, and research initiatives (such as the international doctoral program Re-Thinking Environment: The Ecological Transformation of Society, in the context of which this book emerged), the field took off as a global phenomenon around the turn of this millennium. Those who have since joined the experiment share a critique of the limited focus of earlier environmental histories, its essentialism toward nature and declensionist narratology, and challenge the persistent anthropocentric bias of dominant ethical and political theories. A key and ongoing conceptual point of critique is the nature-culture divide that puts humans outside of and in control of "nature."[27] Importantly, however, Environmental Humanities did not emerge proposing a new object of study, a new humanistic perspective on not-humanistic fields, or a particular set of methods. It emphasized the need to work collaboratively from the humanities and social sciences to adequately respond to the complexity of ecological crises with constructive knowledge and criticism.[28]

As an interdisciplinary field committed to complex modes of analysis, we argue, Environmental Humanities are hence perfectly placed to embrace ambivalence as an approach that recognizes the simultaneity of difference and unity, without leveling global inequality. It can do so in two important ways: first, by fully playing out the field's mindfulness of different methods, approaches, theories, and archives and, second, by placing particular emphasis on its critical ontological approach that seeks to move beyond an understanding of differences to facilitate collaboration and nourish compromise. Everyday practice in the interdisciplinary Environmental Humanities involves not merely describing difference, but also engaging in the ongoing practice of listening to and connecting with the other and finding a joint or compatible methodology, approach, or even solution. The contributions in Part I of this book are living testimony to this process of interdisciplinary collaboration where disciplinary entry points—ranging from vocabulary to citation practices—are introduced and discussed. Contributions in Part II of this book, then demonstrate how scholars from the Environmental Humanities are particularly prone to identifying the inter-connectivity between mental states, economic developments, and socio-ecological transformation processes and the powerful role of ambivalence as both a risk and an opportunity for finding solutions in the face of global ecological threats, collective ignorance, and individual depression. Chapters exemplify how conflicts and perspectives clash between the Global North and the Global South, or how powerful actors engage in greenwashing or condone environmental damage in the pursuit of profit while ignoring the interests of marginalized groups. What comes to the fore are the seemingly irresolvable contradictions between the need for decarbonization and the desire for security and prosperity, along with the question of solidarity and impossible alliances between labor rights and environmental protection. Lived experiences collide with business strategies and the dark sides of taken-for-granted consumerism. What all authors bring to this book helps in maintaining its disciplinary specificity and distinct methodological borders while hoping to cross-pollinize ideas, inspire conversations and new collaborations across the boundaries that exist between academic fields, and secure their respective scholarly standards. This claim positions this book as a contribution to a field that envisions itself as always in the making while creating and maintaining a space for scholarly controversy, interdisciplinary encounters, and fruitful ambivalence.

Conceptually grounded in the Environmental Humanities, and contributing to the self-understanding and the definition of the field, *Ecological Ambivalence, Complexity, and Change* is a multifaceted, highly informed conversation between experts from the humanities, including the humanities-minded social sciences, and economics. The book asks how the humanities *and* the social sciences can work together to help humanity develop and cultivate the individual, collective, and organizational skills that our societies need to overcome paralysis and engage in practical action—all of which needs to be done in a prudent and responsible manner. Each author engages in "writing across the border" of their discipline, thus illustrating a special awareness vis-à-vis related, and less related, fields. This makes this book an exercise in building intellectual pathways between theoretical approaches and case studies, thereby connecting knowledge and insights from anthropology, cultural studies, economics and management research, environmental ethics, history, human geography, literary studies, political science, postcolonial studies, and sociology.[29]

Notes

1 The Nature Conservacy, "Case Study. Belize Blue Bonds for Ocean Conservation," 2023, www.nature.org/content/dam/tnc/nature/en/documents/TNC-Belize-Debt-Conversion-Case-Study.pdf; Naturevest, "Belize Debt Conversion Case Study," last modified May 17, 2022, www.nature.org/en-us/about-us/who-we-are/how-we-work/finance-investing/naturevest/belize-debt-conversion-case-study/.

2 Xiaoyi Jiang and Hao Cao, "Implementing the Debt-for-Nature Swaps for Marine Protected Areas: Case Studies from Seychelles and Belize," *Humanities & Social Sciences Communications* 11, no. 1 (2024): 331–339, https://doi.org/10.1057/s41599-024-02855-3;

Nicholas Owen, "Belize: Swapping Debt for Nature," *IMF Country Focus*, May 4, 2022, www.imf.org/en/News/Articles/2022/05/03/CF-Belize-swapping-debt-for-nature.

3 Jonas Gamso, "Debt-for-Climate Swaps Can Save the Planet. Why Aren't They?" *Stanford Social Innovation Review*, August 15, 2023, https://doi.org/10.48558/RK2K-BS67.

4 Michael Occhiolini, "Debt-for-nature swaps," *Policy Research Working Paper Series* 393 (1990).

5 Georgie Wilson, "Debt-for-nature swaps: The right kind of climate finance or not?" *The Climate Reality Project*, last modified March 13, 2024, www.climaterealityproject.org/blog/debt-nature-swaps-right-kind-climate-finance-or-not; Gamso, "Debt-for-Climate Swaps Can Save the Planet."

6 On the intersectionality of time, space, and body see Simone Müller and May-Brith Ohman-Nielsen (eds.), *Toxic Timescapes. Examining Toxicity across Time and Space* (Athens, OH: Ohio University Press, 2023).

7 Here, the influence of the ontological turn and the works of Bruno Latour, Eduardo Viveiros de Castro, and Philippe Descola are important.

8 See, for instance, Greta Thunberg, "Speech at The U.N. Climate Action Summit," September 23, 2019, Transcript by NPR, www.npr.org/2019/09/23/763452863/transcript-greta-thunbergs-speech-at-the-u-n-climate-action-summit.

9 Red-green colorblindness is a particularly good example to exemplify that people can "color their world" very differently.

10 Zygmunt Bauman, *Modernity and Ambivalence* (Cambridge: Polity Press, 1991), 2.

11 Eugen Bleuler, "Ambivalenz," *Zentralblatt für Psychoanalyse* 1 (1911): 266–268, last modified May 6, 2019, www.sgipt.org/medppp/gesch/ambiv-g.htm.

12 See for instance Ayushi Dhawan and Simone M. Müller (eds.), "Hazardous Hope. Repositioning Troubled Research (and Researchers) in Times of a Troubled Planet," *Environmental Humanities* 16, no. 2 (2024): 433–440, https://doi.org/10.1215/22011919-11150123.

13 Joel A. Tarr, *The Search for the Ultimate Sink: Urban Pollution in Historical Perspective* (Ohio: University of Akron Press, 1996).

14 See, for instance, Christian Jakob, *Endzeit: Die neue Angst vor dem Weltuntergang und der Kampf um unsere Zukunft* (Berlin: Christoph Links Verlag, 2023); Bern Rieken and Reinhold Popp, *Eco-Anxiety – Zukunftsangst und Klimawandel: Interdisziplinäre Zugänge* (Münster: Waxmann, 2021); Joshua Trey Barnett, *Mourning in the Anthropocene: Ecological grief and earthly coexistence (East Lansing:* Michigan State University Press, 2022); Lee Zimmermann, *Trauma and the discourse of climate change: Literature, psychoanalysis and denial* (New York: Routledge, 2020).

15 See Debra J. Davidson, "Rethinking adaptation: Emotions, evolution, and climate change," *Nature and Culture* 13, no. 3 (2018): 378–402; Jordi Xifra and Almiron Núria (eds.), *Climate change denial and public relations: Strategic communication and interest groups in climate inaction* (New York: Routledge, 2020).

16 Michael Hulme, *Why We Disagree on Climate Change. Understanding Controversy, Inaction, and Opportunity* (Cambridge: Cambridge University Press, 2009); Kari Marie Norgaard, *Living in Denial. Climate Change, Emotions, and Everyday Life* (Cambridge, MA: MIT Press, 2011); George Marshall, *Don't even think about it* (New York: Bloomsbury, 2014); Andrew J. Hoffman, *How Culture Shapes the Climate Change Debate* (Stanford: Stanford University Press, 2015); Dale Jamieson, *Reason in a Dark Time. Why the Struggle against Climate Change Failed – And what it means for Our Future* (Oxford: Oxford University Press, 2014).

17 Marlis Charlotte Wullenkord, "From denial of facts to rationalization and avoidance: Ideology, needs, and gender predict the spectrum of climate denial," *Personality and individual differences* 193 (2022).
18 Peter Friederici and Kathleen Dean Moore, *Beyond climate breakdown: envisioning new stories of radical hope* (Cambridge, MA: MIT Press, 2022); Christof Mauch, "Slow Hope. Rethinking Ecologies of Crisis and Fear," *RCC Perspectives: Transformations in Environment and Society* 1 (2019): 1–43, doi.org/10.5282/rcc/8556.
19 In literary studies, for example William Empson's *Seven Types of Ambiguity* (1930), one of the most influential works of literary criticism in the 20th century develops an elaborate system of different kinds of ambiguity for describing the response of art and literature to the complexities of the modern era, thus offering thinking in ambiguity as an alternative to reductionist techno-economic conceptions of modernity.
20 Max Horkheimer and Theodor W. Adorno, Dialektik der Aufklärung. Philosophische Fragmente, in *Max Horkheimer: Gesammelte Schriften,* Band 5, ed. Max Horkheimer (Frankfurt am Main: Fischer, 1987).
21 Ulrich Beck, *World Risk Society: On the Search for Lost Security (Frankfurt am Main:* Suhrkamp, 2007).
22 See Caspar Battegay, "#Ambivalenz: Der Nutzen des Nicht-Entscheidens," *Geschichte der Gegenwart*, last modified January 25, 2023, https://geschichtedergegenwart.ch/ambivalenz-der-nutzen-des-nicht-entscheidens/.
23 Annemarie Mol, *The Body Multiple: Ontology in Medical Practice* (Durham, N.C.: Duke University Press, 2002); Marisol de la Cadena and Mario Blaser, *A World of Many Worlds* (Durham, N.C.: Duke University Press, 2018); Anna Tsing, *Friction: An Ethnography of Global Connection* (Princton, NJ: Princeton University Press, 2005).
24 David Harvey, *Justice, Nature, and the Geography of Difference* (Oxford: Blackwell, 1996); Ramachandra Guha and Joan Martínez-Alier, *Varieties of Environmentalism. Essays North and South* (London: Earthscan, 1997).
25 Battegay, "#Ambivalenz."
26 John Keats, *The Complete Poetical Works and Letters of John Keats* (Boston and New York: Houghton & Mifflin, 1899), 277.
27 William Cronon, "The Trouble with Wilderness or Getting Away with the Wrong Nature," in *Uncommon Ground. Rethinking the Human Place in Nature*, ed. William Cronon (New York: Norton, 1995): 69–90; Sverker Sörlin, "Environmental Humanities. Why Should Biologists Interested in the Environment Take the Humanities Seriously," *BioScience* 62, no. 9 (2012): 788–789; Elizabeth M. DeLoughrey, Jill Didur, and Anthony Carrigan (eds.), *Global Ecologies and the Environmental Humanities: Postcolonial Approaches* (New York: Routledge, 2015); Astrida Neimanis, Cecilia Asberg, and Johan Hedrén, "Four Problems, Four Directions for Environmental Humanities. Towards Critical Posthumanities for the Anthropocene," *Ethics and the Environment* 20, no. 1 (2015): 6797; Deborah Bird Rose et al., "Thinking through the Environment, Unsettling the Humanities," *Environmental Humanities* 1 (2012): 1–5.
28 Ursula Heise, "Planet, Species, Justice—and the Stories We Tell about Them," in *The Routledge Companion to the Environmental Humanities,* ed. Ursula K. Heise, Jon Christensen and Michelle Niemann (London: Routledge, 2017): 1–10; Robert Emmett and David Nye, *The Environmental Humanities: A Critical Introduction* (Boston: MIT Press, 2017).
29 A full survey of the individual contributions can be found in the concluding remarks of this book, authored by Hubert Zapf.

Part I

Conceptual Facets

1 Ecological Transformation and the Implications of Spatial Scale

Matthias Schmidt

Introduction

In view of the consequences of climate change and the global decline in biodiversity, socio-ecological transformation is needed and shared as a normative goal by large sections of society, politics, business, and academia. However, the transformation toward a climate-neutral and environmentally friendly way of life and economy is proceeding very slowly and in small steps, and despite articulated agreement, it often encounters open or hidden resistance or fails due to conscious or unconscious ignorance or lethargy. As such, the necessary radical corrections and measures fail to materialize. This raises questions relating to the reasons for this conscious or, in many cases, unconscious resistance and failure to act. In this chapter, I discuss the ambivalences that go along with ecological transformation and argue for considering the spatial distance and dimensions of the problem in overcoming the current environmental crisis.

Regional differences and spatial distances between the causation of environmentally and climate-damaging activities and their consequences make causalities appear either less obvious or blurred. At the same time, individual contributions to greenhouse gas emissions, resource consumption, and waste production are supposedly insignificant in relation to planetary systems. As a result, individuals experience environmental crises in vastly different ways, thereby relieving polluters of their responsibility.

It is a proven fact that the regions, or better societies, that have contributed significantly to global environmental degradation, climate change, and species extinction differ from those regions or societies that have suffered particularly from the negative consequences of these processes. To hold individuals, groups, or states responsible is rendered difficult by this spatial distance, although causalities can certainly be proven. Added to this is temporal immediacy since climate impacts and environmental damage occur with a considerable time lag from causation. Who is responsible for the damage occurring today if human-induced carbon dioxide content in the atmosphere has accumulated over the past two centuries, the burning of fossil fuels has taken place over several generations, or large-scale deforestation and the draining of peatlands took place decades ago?

DOI: 10.4324/9781032627984-3

At the same time, individual behavioral changes, or restrictions to act in the most environmentally and climate-friendly way possible, seem insignificant or even irrelevant in the face of the inertia of climate processes and unstoppable climate change—positive feedback is not immediately noticeable. Questions on balancing different interests, gains, and costs also contribute to inaction or transformative lethargy: ecological concerns are pitted against economic interests, questions of societal burden and cost-sharing remain unresolved, and numerous dilemmas, such as intra-ecological conflicts between climate protection and nature conservation, make clear answers difficult.

Among many other economic, political, cultural, societal, or psychological reasons for the apparent ambivalence toward the ecological transformation considered necessary by large segments of society, spatial distance and dimensions are often too little considered. I argue, then, that distance and dimension play a not insignificant role in the reluctance to engage in ecological transformation; they invite people to fade out, overlook, and ignore.

The (Neo)Colonialism of Climate Change

While among stratigraphers and geologists the debate about the beginning of the Anthropocene oscillates between the Neolithic Revolution and the peak of the radioactive fallout in the mid-1960s,[1] the causation of global climate change can be narrowed down much more temporally. The conversion of the energy system from solar-driven energy sources such as hydropower, wind, and wood to the use of fossil fuels in the course of the Industrial Revolution was the relevant starting point for the disproportionate and ongoing increase in global carbon dioxide emissions.

Starting in England as the mother country of industrialization at the end of the 18th century, it was especially European countries and later North America, as well as Japan and Australia, that were responsible for the bulk of emitted carbon dioxide. The increased proportion of the gas in the atmosphere caused by humans thus essentially represents an accumulation of combustion processes that have occurred over the course of the last two centuries. Similarly, the emission rates of more recent years also demonstrate very clearly the disproportionately large role of the Western world. Although, in a comparison of countries, China is now the largest emitter of greenhouse gases, the share of the so-called "Global North" (Europe, North America, Japan, Australia) is still 34.8%,[2] which is clearly disproportionate in view of its share of the world population of 18.7%.

This inequality in the causation of climate change corresponds almost cynically with the unequal impact of its effects. Although droughts, heatwaves, water shortages, rising sea levels, and extreme weather events are now increasingly affecting the Global North, as recently demonstrated by the searing heat and devastating fires in North America and the Mediterranean region, as well as various flood and storm events in Europe, exposure to and, above all, the destructive consequences of the extreme conditions are significantly greater in the Global South. Exceptional heatwaves in India, monsoon floods in Pakistan, extreme droughts in the Horn of

Africa, threats from rising sea levels to island states in the Pacific, increases in the number and intensity of typhoons in Southeast Asia, or hurricanes in the Caribbean affect hundreds of millions of people in highly vulnerable societies.[3]

The functional link between the Global North and the Global South, which has existed at least since colonialism and is based on unequal power relations, is perpetuated in the current environmental and climate crisis. Countries and the people of Africa, Asia, and Latin America were seen by colonial powers as resource depots and labor pools, and for centuries, they have served the wealthy states as cheap production sites or as disposal areas for pollutants[4] and, most recently, as compensation for their own climate and ecological sins as carbon sinks and the last remaining biodiversity hotspots.

Due to the increasingly alarming reports published by the Intergovernmental Panel on Climate Change (IPCC),[5] as well as a multitude of natural disasters caused by climate change, numerous governments and large parts of societies worldwide are stirred up and recognize the necessity for a socio-ecological turnaround—a conversion of the energy system, the economy, and consumption patterns—and the ultimate objective of climate neutrality. This desire is expressed primarily in prominently presented—but almost unrealistic—goals, such as making one's own economy climate-neutral by a certain year (Germany by 2045, EU and USA by 2050, and China by 2060), as well as in new laws and measures, bans, and subsidies. Significantly, however, global inequality also repeats itself in these measures: in achieving its goals on the level of nation states, the Global North involves the Global South as a helper on its way to climate neutrality, ascribing a role to these countries just to make sure it maintains its own prosperity. The following examples exemplify this notion in more detail.

Trading in CO_2 certificates explicitly involves the less wealthy countries. In the sense of indulgence trading, companies in the Global North can initiate carbon reduction projects in the Global South so that they do not have to reduce their own greenhouse gas emissions. This can take the form of promoting energy- and emission-saving technologies, preserving forests and peatlands, or initiating projects for renaturation or reforestation.[6] Such measures turn the Global South into the carbon sink of the Global North, and while these activities help construe a "greener" image for carbon-intensive companies and societies, the success or dubiousness of this trade is controversially debated in science and politics.[7]

Over the course of the past two millennia, the forest areas of Europe and other regions of the Global North have been massively reduced. Areas untouched by humans exist only sporadically, and biodiversity has declined sharply. This explains why efforts to protect and preserve wilderness and biodiversity are directed at the remaining hotspots in tropical and subtropical regions. This is often done through the intervention of international organizations and NGOs, or by corporations or committed individuals from wealthy northern nations. Often, they do so without considering the concerns of local populations, which is why the practice is also referred to as "green grabbing,"[8] i.e., global concerns for species protection and climate policy being placed above the local or national concerns of societies in the Global South.

For decades, numerous pollutants and waste generated by industrial processes or consumption have been dumped by wealthy economies in the Global South. Discarded smartphones, computers and consumer electronics, batteries, and wrecked tankers end up in landfills in West Africa or South Asia, where they are eviscerated, burnt, or dumped at considerable environmental and health risks.[9] Thanks to this international trade in waste, the Global North not only gets rid of its pollutants, but it also achieves largely clean rivers and tolerable air quality in its own rural and even urban regions.

The above examples support the thesis proffered by sociologist Stephan Lessenich,[10] who describes the advanced political economies of the Global North as "externalization societies," whose wealthy majority benefit from the resources, labor, and limited life opportunities of others: "The Global North can live, produce, and consume as it does because others do not—and cannot—do so."[11] Brand and Wissen[12] denounce the "imperial way of life" of the Global North, according to which consumerism and the maintenance of prosperity in Europe and North America, but also by global elites in general, take place massively at the expense of underprivileged social groups, especially in the Global South. In this respect, ecological transformation, the rules and direction of which, moreover, are largely prescribed by the Global North, can thus also be seen as a neocolonial project.

There is no doubt that climate change, species extinction, and the negative consequences for the vulnerable half of the world are being taken seriously by many politicians and societies in the Global North. Many governments of advanced political economies in the North also acknowledge their responsibility, and their economies are making a supposedly greater contribution to overcoming the environmental and climate crisis, albeit essentially in the form of technological restructuring and financial expenditure, rather than in the form of relinquishing power and influence, or restricting cherished lifestyles, or genuinely sharing the burden, for example, by taking in environmental refugees. Indeed, the interests of the country they represent are always at the forefront of the political actions of state governments, and the fear of imposing too much on their populations is considerable. As recent events and developments demonstrate, almost all concrete policy measures to reduce greenhouse gases, if they involve higher costs or restrictions, provoke resistance. In the German context, for example, this can be seen in public protests against increasing taxes on fuels, the expansion of wind-power farms[13] or the current dispute on the banning of gas and oil heating systems. The public's widespread resistance or unwillingness to change prevents political parties in the Global North from intervening in the status quo, which leads to a deceleration of the necessary socio-ecological transformation. As a result, the global environmental and climate crisis perpetuates existing inequalities, and those who have contributed the least pay the bill.

It's the Geography—Regional Differences and Spatial Distances

The cumulative problems of environmental degradation, species extinction, and climate change have reached a global dimension and invite taking a planetary perspective, which is also reflected in the growing importance of respective academic disciplines such as Earth System Sciences,[14] Planetary Health,[15] or Planetary Histories.[16] Nevertheless, it should not be forgotten that the Earth is first and foremost a planet that, due to its heterogeneous surface consisting of an irregular mosaic of oceanic and terrestrial areas, has an almost infinite number of different site conditions and is thus characterized by the greatest heterogeneity. Without this diversity, there would perhaps be no need for scientific geography, which as a discipline deals with the processes and structures of the geosphere in a wide variety of places, sites, or regions. Topographical position (location, altitude), solar radiation, geology, hydrology, and consequently climatic and ecological conditions, as well as millions of years of natural history and thousands of years of human history, result in the most diverse living conditions worldwide.

But the world has shrunk dramatically in the last two centuries due to modern technology,[17] and it is now often referred to as a "global village."[18] Furthermore, spatial distances can be covered faster and more easily, the possibilities for mobility and exploration have grown significantly, and worldwide communication is possible in real time. In view of technological developments and expanded communication services, as well as in relation to leveling and homogenizing processes of the present—often termed "McDonaldization,"[19] the "end of geography"[20] has occasionally already been anticipated. Nevertheless, even in times of globalization, spatial distances or the lack of larger spatial distances that allow for direct inspection or face-to-face communication, questions of locality, and a sense of place and of identity in a place continue to play a very important role for social processes and individual actions. This also manifests in the so-called "Spatial Turn" in the social sciences and other academic fields,[21] according to which the importance of space for understanding social processes is emphasized.

Geography, and especially economic geography, has dealt with questions of spatial location, optimal site conditions, and site relations for many decades, for example, to reduce transport times and costs (e.g., Walter Christaller, August Loesch, and Peter Haggett),[22] with a focus on spatial distances between production sites, raw material deposits, and consumers. Automobile industries in the Stuttgart area in Southern Germany, for example, where numerous supplier and logistics companies have settled alongside actual car companies, can serve as an illustration in this regard. While in today's information technology or innovation clusters, transport costs are of little importance, short spatial distances contribute to the success of knowledge-based innovation regions. Silicon Valley is a prime example of this point, in that the spatial proximity between major enterprises, startups, and universities is an important success factor that attracts highly qualified specialists.

The extent to which one's own microcosm dominates everyday life, even in highly developed and technically highly networked societies in the century, is also revealed when we look at rural areas and media landscapes in the USA, whereby local and regional news dominate and ignorance about other regions is frighteningly high ("What I don't know won't hurt me"). The increasing importance of social media, which technically makes it possible to find out about events in the most diverse and distant places around the world, does not necessarily lead to a spatial-mental broadening of horizons—a notion supported by the fact that the algorithms that react to our smartphone use direct our attention to objects within our own bubble.

Remote spaces and their characteristics, structures, and problems are still perceived by most people as being far away and largely irrelevant,[23] especially if they do not seem to affect their own lives. The famous sack of rice falling over in China is of no interest. In addition, of course, there is the extent to which one is affected; if the problems do not affect me, then they are of no interest to me ("What do I care about the concerns of others? Everyone is closest to himself."). But even when spatially distant problems or crisis events are consciously perceived, such as a drought or crop failures in the U.S. Midwest, which have actual consequences for rising corn prices worldwide, people usually find it difficult to accept the causalities and to draw appropriate conclusions. These are all too human and well known characteristics,[24] but to consider seriously the consequences of ecological crises that occur at a seemingly safe distance away from home, and to move forward in implementing and speeding up necessary ecological transformation, remains one of the most pressing yet challenging issues and tasks of our era, so that people can be moved or motivate themselves to act against the abstract, often only indirectly perceptible phenomenon of climate change.

The Spatial Dimension of Reluctance

Many factors contribute to the resistance or unwillingness to act in relation to the ecological transformation of society on a scale that meets the enormity of the challenge. In this chapter, I limit myself to discussing those that relate to spatial distance and spatial dimension.

Back to the National and Regional

After a heyday of globalization, including a positively connotated discourse on globalization, whose economic advantages and mobility gains stood in the foreground for a long time, the pendulum has started to swing in the other direction for the last few decades, namely that the narrative of a globalization that benefits everyone is losing its persuasive power.[25] Instead, fear of globalization and a return to national or regional spaces now dominate the discourse. Fueled by global economic crises, nationalist policies (e.g., "America First," "Hindutva"), warlike conflicts, and policies that focus on the national or even the regional and provincial are gaining adherents and labeling anything beyond one's geographical borders as "foreign" or even "hostile and damaging." Additionally, egocentrism and navel-gazing are on the rise, as are isolation and ignorance. In view of proven global problems such as climate change and biodiversity loss, which urgently require joint global action, such attitudes and national go-it-alone approaches are problematic or counterproductive, to say the least.

It appears that the concept of globalization and its practical implementation have been approved by the dominant states of the North as long as they can profit from it. This includes, for example, the importation of goods and services from regions where they can be produced more cheaply, or the disposal of waste or the discharge of pollutants into the global system; in these cases, the global division of labor and tasks is highly welcome. If, however, one's own profits are diminished, for example, by having to bear the costs of the problems and damages caused by one's own prosperity, nation states (of the Global North) try to withdraw from their global responsibility and align their policies nationally. In addition, undoubtedly, disruptions to supply chains, triggered by armed conflicts, pandemics or political confrontations, or concerns about one-sided dependencies, have led to an increasingly critical view of the concept of globalization. This reduced focus on national, regional, and local concerns clearly has a spatial component and contributes to the ambivalence toward ecological transformation.

Dimension of Climate Change and Its Consequences

Human-made climate change is characterized by steadiness and relative slowness, by a time lag between causation and consequences, and by a planetary dimension. Global emissions of greenhouse gases causing climate change have been increasing more or less continuously since the Industrial Revolution, but there are significant regional differences worldwide with respect to sources of origin. Although the connections between the anthropogenically increased emissions of greenhouse gases and changes in the climate have long been scientifically proven, causalities of concrete (emission) events and their climate consequences cannot be directly derived or proven. This issue has been used by climate change deniers to build an argument against even the most undisputed causality between anthropogenic greenhouse gases released into the atmosphere and climate change.

The consequences of climate change are also temporally delayed and spatially dispersed; furthermore, they often occur randomly and abruptly and thus cannot be precisely predicted, either temporally or spatially. On the one hand, this vagueness has an unsettling effect, but on the other hand it is also reassuring and relieving, since weather- or climate-related destructive events do not necessarily have to affect oneself and the own environment at a close point in time.

In addition, there is the imperceptible (self-)effectiveness of one's own actions or changes in actions toward climate- and environmentally friendly consumption ("Can I change anything at all?"), for example, by traveling less by air, consuming less meat, or heating in an energy-saving manner. One's own behavioral changes show no tangible results, no rewards; thus, the immediate causality that people otherwise experience in many of their actions is missing. In addition, one's own contribution to causing climate change, which can thus be classified as insignificant, may lead people not to reduce it further or to cease their efforts altogether. In many cases, the dimension of climate change that goes beyond one's own comprehension and imagination also leads to intimidation and uncertainty, to a feeling of helplessness—and thus often to the rigidity of shock, just like the famous rabbit caught in the headlights.

Dimension of the Commons

The difficulty in implementing a socio-ecological transformation is also rooted in the dimension of the group involved—the global community. Work on common pool regimes[26] demonstrates the ability of groups to manage and use common goods together successfully and quite sustainably; the atmosphere can also be considered a global common good. However, success requires functioning, flexible institutions recognized by all, as well as a degree of mutual trust,[27] but neither of these is present when it comes to anthropogenic climate change. For one thing, there is a lack of appropriate institutions, for even if international commitment agreements are concluded with the signatures of state leaders, effective sanction mechanisms are lacking. Moreover, there is no common political authority recognized by all states whose requirements will be followed. Second, trust among the various states is rather limited, since they either belong to different political blocs or give priority to the interests and goals of their own states.

The global climate and environmental crisis in particular can only be dealt with effectively through cooperation and coordination, but it is often responded to with individual rather than collective approaches and solutions.[28] This can be explained by the fear of making individual losses when investing in cooperative solution attempts, by distrust of other group members, or by a lack of belief that cooperative strategies are promising. Particularly in recent decades, innovations, technological developments, and changes in economic structures have transformed collective problem-solving approaches from a necessity to a possibility. This in turn has reinforced the tendency to favor individualistic strategies.[29]

Significance of Climate and Environmental Issues

The consequences of climate change, species extinction, and other environmental problems are now recognized in politics, science, and society as global—even existential—problems. Nevertheless, for a wide variety of reasons, they are not equally regarded as priorities by all people. Underprivileged societies or individuals who have to fight for their daily survival, for example, to ensure their own nutrition, who suffer from disease or violence, or who are affected by other deficiencies or processes affecting their own well-being are not able to care about avoiding the climate or environmental consequences of their actions. Not to mention the fact that the less affluent population groups contribute least to the negative climate impacts. Warlike conflicts and fear of persecution or oppression are also existential threats and push environmental and climate issues into the background. However, aspects that are less existential but do affect people's own prosperity, well-being, or social status, such as concerns about economic and social decline, often lead people to attach less importance to the environmental and climate crisis.

The Global Risk Report 2023,[30] based on assessments by 1,200 experts and political and business leaders, identifies the most important short-term risks as the cost-of-living crisis, natural disasters, extreme weather events, geoeconomic confrontation, failure to mitigate climate change, the erosion of social cohesion, and societal polarization. Two of these are environmental issues. Among long-term risks, environmental issues rank in the first four places: failure to mitigate climate change, failure to ensure climate-change adaptation, natural disasters and extreme weather events, and biodiversity loss and ecosystem collapse. This assessment of global risks clearly shows the urgency and existential nature of today's environmental problems.

However, a look at the problems and concerns of people in an industrialized country like Germany reveals a different picture. In a 2023 survey asking about the biggest problems currently facing the nation, 47% named rising prices/inflation/cost of living, 24% environment and climate change, 24% immigration, 19% energy supply, and 15% housing construction/procurement.[31] Although respondents in this and similar surveys have given high priority to climate and environmental issues for several years, they sit far behind immediate financial concerns. Moreover, it remains unclear as to what extent these concerns also relate to negative climate change impacts on people outside their own country. Rather, by naming immigration as one of the biggest problems, it is more likely to assume that respondents tend to emphasize concerns about their own life, security, and prosperity. The spatial distance to climate and environmental problems is also clearly expressed here. The consequences of climate change in the Global South play no role, or at most a subordinate role, as perceived concerns in the Global North. Rather, attempts are made to secure one's own prosperity despite the obvious—sometimes catastrophic—consequences of the environmental and climate crisis on other, foreign people, or even at their own expense, by erecting fences and walls to protect themselves from potential consequential problems. This, too, has a spatial dimension, for an inside and an outside are created, i.e., a safe space versus an unsafe space.

Attention to and empathy for problems in the Global South appear to be subject to considerable fluctuations in affluent societies. This was made clear in an almost cynical manner by two boat-related accidents occurring in quick succession in June 2023: The accident involving the submarine "Titan" off the coast of Canada with five dead—exclusively very wealthy people who had voluntarily joined this exploration tour to the wreck of the Titanic—attracted considerably more attention in the media worldwide than the capsizing of a refugee boat in the Mediterranean Sea with presumably more than five hundred casualties, each of whom had squeezed into an unseaworthy boat to escape existential threats in their home countries.

Inaction and Psychological Distance

Ecological transformation requires multifaceted efforts at all levels, in all sectors, and by all groups; a mix of prohibition and renunciation appears necessary. Whereas in the articulation of regulations, politics is called upon to pass appropriate laws that hold industry, municipalities, and individual citizens accountable, whereby bans explicitly named as such are hardly communicable in democratic societies, in the aspect of renunciation, responsibility is passed on to individuals, who often find themselves in a dilemma, as renunciation or changes in behavior to achieve greater goals are often difficult. The German writer Kurt Tucholsky noted in this regard that "we ought to, but we don't do." This is where Richard Gifford's seven "dragons of inaction"[32] come in, which contribute to why we fail to act appropriately in an environmentally or a climate-friendly way despite knowing better.[33] Specifically, the reasons for our inaction are our limited thinking capacity, ideology, comparisons with others, irreversible costs, disapproval, perceived risks, and limited action. With regard to the argumentation of this chapter (spatial dimension), limited thinking capacity or psychological distance are especially relevant, according to which our brain is strongly oriented toward living in the here and now.

Psychological distance leads to perceiving climate change and its impacts as abstract and distant in space and time.[34] Even defining climate as a 30-year average of weather patterns requires a certain degree of abstraction.[35] More significantly, most of the consequences of anthropogenic climate change will only be visible in the future, so they are thus distant in time,[36] which results in a public perception that climate change is more relevant to distant regions and times than to the here and now.[37] The Construal Level Theory[38] shows how distances influence perceptions, decisions, and behavior. It argues that the subjective psychological distance at which people perceive events or objects determines how concretely or abstractly these entities are mentally represented and might entail corresponding consequences for action.[39] In addition, there is the importance of place attachment: if I have a strong attachment to the place where I live, I am more likely to care about it. The ecological catastrophes in the Global South, on the other hand, tend to leave me cold.

This spatial and temporal distance from the consequences of the environmental and climate crisis plays into the hands of another dragon of inaction,[40] namely disapproval and denial, and because the mostly distant consequences are usually conveyed through the media, they are easier to deny or minimize ("This can't be happening at all").

Another dragon of inaction related to spatial distance is the comparison with other people that influences one's own actions.[41] This is because people compare themselves with others in their environment and not with people from remote world regions. If their neighbors fly twice a year on vacation, then the reduction down to only one flight per year in relation to this already represents a perceived restriction that can be easily explained. The fact that the majority of people worldwide never take a plane trip in their lives is irrelevant for such a decision and justification, or it is ignored.

Conclusion

The socio-ecological transformation desired and demanded by many people worldwide is taking place, but much too slowly for some, too quickly for others. Despite its necessity, justified by science, it is questioned, evaded, ignored, or fought against. There are numerous reasons and studies that explain this seeming contradiction that leads to a discrepancy between knowledge and action. In this chapter, I have tried to demonstrate that the inertia of ecological transformation, discomfort, or fading out is also closely related to spatial and temporal distances and dimensions. I have argued that spatial distances to the media-mediated hotspots of climate change, as well as the dimension of the environmental and climate crisis, contribute to the failure to change actions, or to changing them only slightly.

In view of the widespread knowledge of the environmental crisis, the attitudes and behaviors of consumers in the Global North, as well as decision makers in politics, business, and society, may well be described as "willful blindness."[42] This willful blindness can result in several phenomena and behaviors, such as "turning a blind eye" to justify particularly climate-damaging behavior, "burying one's head in the sand" in order not to perceive unpleasant facts such as increasing droughts or extreme weather events, "denying reality" with regard to one's own threat from, for example, rising sea levels, "selective awareness" by ignoring the environmental and climate impacts of one's own lifestyle, and "consciously avoiding" knowledge about poor working conditions in the production of one's own consumer goods, for example, in the extraction of cobalt in Congo. Willful blindness can thus also be seen as a more calculated and strategic form of ignorance.

Even though the probability of being affected by climate change-related catastrophes or unfavorable developments is increasing, most events will continue to take place at great spatial distance. Climate change will continue to progress in the coming decades, and even radical behavioral changes will not alter the situation. Nor can the scale of the problem be changed. What can change, however, is knowledge, attitudes, and perceptions. To do this, it is necessary to increase understanding of the interrelationships with and empathy for those most affected. This is where the Environmental Humanities can make an important contribution, because in addition to scientific facts and forecasts, which can certainly lead to fatigue and jadedness, dealing with this global problem is also very much related to social and cultural factors.

Notes

1 Jan Zalasiewicz et al., *The Anthropocene as a Geological Time Unit: A Guide to the Scientific Evidence and Current Debate* (Cambridge: Cambridge University Press, 2019). After the Anthropocene Working Group defined in July 2023 the beginning of the Anthropocene as 1950 and used the sediments of Lake Crawford in Canada as a reference, the International Commission on Stratigraphy surprisingly rejected the proposal in March 2024 of a new Anthropocene geological epoch. Christoph Antweiler, *Anthropologie im Anthropozän: Theoriebausteine für das 21. Jahrhundert* (Darmstadt: Wissenschaftliche Buchgesellschaft, 2022).

2 Hannah Ritchie and Max Roser, "CO_2 emissions," *Our World in Data*, accessed July 2, 2023, https://ourworldindata.org/co2-emissions#co2-emissions-by-region.

3 Hans-Otto Pörtner et al. (eds.), *Climate Change 2022: Impacts, Adaptation and Vulnerability. Contribution of Working Group II to the Sixth Assessment Report of the Intergovernmental Panel on Climate Change* (Cambridge: Cambridge University Press, 2022).

4 Simone Müller, *The Toxic Ship: The Voyage of the Khian Sea and the Global Waste Trade* (Washington: Weyerhaeuser Environmental Books, 2023).

5 Pörtner, *Climate Change 2022*.

6 Jeremiás Máté Balogh, "The Impacts of Agricultural Development and Trade on CO_2 Emissions? Evidence from the Non-European Union Countries," *Environmental Science and Policy* 137 (2022): 99–108.

7 Benjamin K. Sovacool, "Four Problems with Global Carbon Markets: A Critical Review," *Energy & Environment* 22, no. 6 (August 2011): 681–694. Raphael Calel, "Carbon markets: a historical overview," *WIREs Climate Change* 4, no. 2 (2013): 107–119.

8 James Fairhead, Melissa Leach and Ian Scoones, "Green Grabbing: A New Appropriation of Nature?" *Journal of Peasant Studies* 39, no. 2 (April 2012): 237–261. DOI: 10.1080/03066150.2012.671770.

9 Simone Müller, "Hidden Externalities: The Globalization of Hazardous Waste," *Business History Review* 93, no. 1 (2019): 51–74. DOI: 10.1017/S0007680519000357.

10 Stephan Lessenich, "The Externalization Society: Living Beyond the Means of Others," *The Futures We Want. Global Sociology and the Struggles for a Better World*, September 25, 2015, accessed August 29, 2023, http://futureswewant.net/stephan-lessenich-externalization-society/.

11 Stephan Lessenich, *Neben uns die Sintflut. Die Externalisierungsgesellschaft und ihr Preis* (Berlin: Hanser, 2016).

12 Ulrich Brand and Markus Wissen, *Imperiale Lebensweise. Zur Ausbeutung von Mensch und Natur in Zeiten des globalen Kapitalismus* (München: Oekom, 2017).

13 Stephan Bosch and Matthias Schmidt, "Wonderland of Technology? How Energy Landscapes Reveal Inequalities and Injustices of the German Energiewende," *Energy Research & Social Science* 70 (2020): 101733. DOI: 10.1016/j.erss.2020.101733.

14 Will Steffen et al., "The Emergence and Evolution of Earth System Science," *Nature Reviews Earth & Environment* 1 (January 2020): 54–63. DOI: 10.1038/s43017-019-0005-6.

15 Samuel Myers and Howard Frumkin (eds.), *Planetary Health: Protecting Nature to Protect Ourselves* (Washington and Covelo: Island Press, 2020). DOI: 10.5822/978-1-61091-966-1.

16 Dipesh Chakrabarty, *The Climate of History in a Planetary Age* (Chicago: The University of Chicago Press, 2021).

17 David Harvey, *The Condition of Postmodernity* (Oxford: Basil Blackwell, 1989). Scott Kirsch, "The Incredible Shrinking World? Technology and the Production of Space," *Environment and Planning D: Society and Space* 13 (1995): 529–555.

18 Marshall McLuhan, *The Gutenberg Galaxy: The Making of Typographic Man* (Toronto: University of Toronto Press, 1962).

19 George Ritzer, *The McDonaldization of Society* (Los Angeles: Pine Forge Press, 2009).

20 Richard O'Brien, *Global Financial Integration: The End of Geography* (New York: Foreign Relations Press, 1992).

21 Charles W. J. Withers, "Place and the 'Spatial Turn' in Geography and in History," *Journal of the History of Ideas* 70, no. 4 (October 2009): 637–658.

22 Daoqin Tong and Alan T. Murray, "Spatial Optimization in Geography," *Annals of the Association of American Geographers* 102, no. 6, (2012): 1290–1309. DOI: 10.1080/00045608.2012.685044.

23 Dragos Simandan, "Proximity, Subjectivity, and Space: Rethinking Distance in Human Geography," *Geoforum* 75 (2016): 249–252. DOI: 10.1016/j.geoforum.2016.07.018.

24 Robert Gifford, "The Dragons of Inaction: Psychological Barriers that Limit Climare Change Mitigation and Adaptation," *American Psychologist* 66, no. 4 (2011): 290–302. DOI: 10.1037/a0023566.

25 Joseph E. Stiglitz, *Globalization and Its Discontents* (New York: Norton, 2002).

26 Elinor Ostrom, *Governing the Commons: The Evolution of Institutions for Collective Action* (Cambridge: Cambridge University Press, 1990).
27 Ostrom, *Governing the Commons*.
28 Jörg Gross and Carsten K. W. De Dreu, "Individual Solutions to Shared Problems Create a Modern Tragedy of the Commons," *Science Advances* 5, no. 4 (April 2019). DOI: 10.1126/sciadv.aau7.
29 Gross and De Dreu, "Individual Solutions."
Katharina van Bronswijk, Jan-Ole Komm and Ingo Zobel, "Die Evolution der Drachen der Untätigkeit: Aktuelle Erkenntnisse zur Lücke zwischen Wissen und Handeln," in *Climate Action – Psychologie der Klimakrise. Handlungshemmnisse und Handlungsmöglichkeiten*, ed. Lea Dohm, Felix Peter and Katharina van Bronswijk (Gießen: Psychosozial-Verlag, 2021), 112.
30 "The Global Risks Report 2023," *World Economic Forum*, January 2023, accessed June 15, 2023, www.weforum.org/reports/global-risks-report-2023/.
31 Europäische Kommission, "Umfrage zu den wichtigsten Problemen für Deutschland 2023," *Statista Research Department*, last modified July 13, 2023, https://de.statista.com/statistik/daten/studie/2739/umfrage/ansicht-zu-den-wichtigsten-problemen-deutschlands/.
32 Gifford, "The Dragons of Inaction."
33 Bronswijk, Komm, and Zobel, *Die Evolution Drachen*.
34 Adrian Brügger, "Understanding the Psychological Distance of Climate Change: The Limitations of Construal Level Theory and Suggestions for Alternative Theoretical Perspectives," *Global Environmental Change* 60 (2020). DOI: 10.1016/j.gloenvcha.2019.102023.
35 Pörtner, *Climate Change 2022*.
Elke U. Weber, "Experience-Based and Description-Based Perceptions of Long-term Risk: Why Global Warming does not Scare Us (Yet)," *Climatic Change* 77, no. 1–2 (July 2006): 103–120. DOI: 10.1007/s10584-006-9060-3.
36 Gifford, "The Dragons of Inaction."
Weber, "Experience-Based."
Sabine Pahl et al., "Perceptions of Time in Relation to Climate Change," *Wiley Interdisciplinary Reviews: Climate Change* 5, no. 3 (2014): 375–388. DOI: 10.1002/wcc.272.
37 Scott Hanson-Easey et al., "Speaking of Climate Change: A Discursive Analysis of Lay Understandings," *Science Communication* 37, no. 2 (February 2015): 217–239. DOI: 10.1177/1075547014568418.
Zoe Leviston, Jennifer Price and Brian Bishop, "Imagining Climate Change: The Role of Implicit Associations and Affective Psychological Distancing in Climate Change Responses," *European Journal of Social Psychology* 44, no. 5 (July 2014): 441–454. DOI: 10.1002/ejsp.2050.
38 Yaacov Trope and Nira Liberman, "Construal Level Theory," in *Handbook of Theories of Social Psychology*, ed. Paul A. M. van Lange, Arie W. Kruglanski and E. Tory Higgins (Thousand Oaks: Sage, 2012), 118–134.
39 Brügger, "Understanding the Psychological Distance."
40 Gifford, "The Dragons of Inaction."
41 Ibid.
42 Judith Bovensiepen and Mathijs Pelkmans, "Dynamics of Wilful Blindness: An Introduction," *Critique of Anthropology* 40, no. 4 (October 2020): 387–402. DOI: 10.1177/0308275X20959432.

2 Climate Solidarity, Time, and Ambivalence

On a New Term and Its Historical Legacy

Dietmar Süß

Introduction

"Humanity has one choice: cooperate or perish." In his opening speech at the 2022 World Climate Summit in Sharm el-Sheikh, Egypt, UN Secretary-General António Guterres made a strong call for a new global climate alliance: "It is either a Climate Solidarity Pact—or a Collective Suicide Pact."[1] Developments at the conference, however, very quickly made clear how difficult it can be to determine what exactly is meant by the term "climate solidarity." In the commemorative United Nations calendar, solidarity has been given its own space, and in this regard, since 2005, the United Nations (UN) has celebrated the "International Day of Human Solidarity" every year on December 20, just before Christmas. The UN proclaims "unity in diversity," and the day is intended to remind people to comply with international agreements, to stimulate new initiatives to fight poverty in the world, and, more generally, "to launch a debate on how to promote solidarity towards the achievement of the Sustainable Development Goals."[2]

The International Day of Human Solidarity is a remnant of the debates about the UN's Millennium Declaration of 2000, and solidarity has become a key expression for worldwide responsibility to eliminate global inequalities. The UN's self-understanding is based on the idea that "those who either suffer or benefit the least deserve help of those who benefit the most. Consequently, in the context of globalization and the challenge of growing inequality, strengthening international solidarity is indispensable."

At the time of its inception, international conflicts, it was assumed, could only be resolved through cooperation and practiced solidarity. For a day so close to Christmas, these were a lot of tasks to deal with at once. Furthermore, in 2000, the UN's Millennium Declaration did not yet include the idea of "climate solidarity," but it has been one of the new key concepts of global environmental movements since the 2010s. UN Secretary-General Guterres made the term popular in March 2023 in his call for a Climate Solidarity Pact. In the political language of climate diplomacy, the terms "climate solidarity" and "climate justice" are often used interchangeably; however, the idea of the former example refers more strongly to the social connectedness of the various actors and political traditions of social movements. Moreover, it connects older traditions of anti-globalization movements

DOI: 10.4324/9781032627984-4

with the protest forms of a newer environmental movement, such as Fridays for Future or Extinction Rebellion, that contextualize ecological and social issues of the 21st century.

In controversies relating to climate solidarity, the reduction of climate-damaging emissions, the financing of "green" investments, access to markets, or the use of natural resources, it often remains unclear what exactly is meant by "solidarity," who is included, who is excluded, and to what extent its interactions (may) result from the climate crisis. Who are the winners, and who are the losers in socioecological transformation? In terms of historical explanation, we note the idea that there can—and must—be something like global climate solidarity. What seems self-evident from the perspective of social movement activists, however, is less clear from a historical perspective, as the different uses of the concept of solidarity do not only reflect various historical contexts or normative foundations. They are also two facts closely related, for solidarity is both a political term of contention and a category of scientific description. Sometimes, the term refers to a type of action, sometimes to a conception of value and utopia, sometimes to a mode of social integration, and it is in these different evocations and their occasional mismatching that ambivalences of ecological transformations find their expression.

The term "climate solidarity" reflects these different levels of meaning and is an expression of the pluralization of solidarity semantics that has been evident since the 1970s. Spoken in the mode of climate solidarity, it is possible—especially for previous social movements such as the labor movement—to find a new connection to the debates about the ecological transformation of our present. Furthermore, it makes it easier to emphasize class-specific lines of conflict embedded in environmental catastrophe.

The history (and political practice) of solidarity is full of contradictions and ambivalences and is closely linked to the history of the development of nation states in the 19th century and the unfolding of modern welfare states and different forms of social integration. At the same time, the idea of solidarity within the labor movement always referred to matters beyond one's own national borders, but its history is by no means romantic, either, since solidarity is not only a "prosocial behavior,"[3] but is also shaped by different lines of tension. An example: one of the most famous and style-defining depictions of solidarity is the wood engraving by the British artist Walter Crane from 1890. Crane was himself part of the still young socialist movement, and in his drawing, which was soon widely used on both sides, he sketched the image of a new social order. In it, the spirit of solidarity made the (male) workers from all continents dance peacefully around the globe, protected by a female angel of freedom. This solidarity was not bound by borders, and it united workers from all nations and continents. Solidarity relations, as the early labor movement thought of them, created new social networks in a special way, in that it changed political spaces and was based on the feeling of common dependencies. At the same time, as this new design also told us, the wood engraving showed only male workers, and the only woman was a floating angel—not an exploited seamstress or a domestic worker. Solidarity practices excluded parts of one's own class and could create new, sexist, and even racist hierarchies. The history of solidarity is thus always a history of inclusion and exclusion, of the tensions between universal and particular norms.

The following reflections tie in with this basic consideration. The article extrapolates the ambivalences inherent in today's notion of (global) climate solidarity, as they are based not only on the plurality of different notions of solidarity as a political evocation and a social practice but also on a contentious relationship between social democratic thought and modern environmentalism. The article commences by retracing—reflecting on the West German debates since the 1970s—the history of solidarity ideas, before teasing out lines of conflict between the newly emerging environmental movement and Germany's established Social Democrats. Expanding the perspective to include German intellectuals' engagement with what were then called "Third World" countries, the article also includes debates about "alternative economies" and the future of Fordist forms of production, as they have also been discussed within the political left since the 1980s. Ultimately, this contribution argues, it is the debate on notions of progress and growth that has decisively shaped the current thought processes on climate solidarity and its embedded ambivalences.

It is no coincidence that the article focuses on the West German debates. In no other European country is there currently a Green-led Ministry of Economics that formulates such a comprehensive claim for a socioecological transformation policy. Naturally, there is always a gap between aspirations and reality, but in order to understand this particular political constellation of a "traffic light coalition," it makes sense to clarify some of the historical depths of the controversies around the idea of solidarity, which is directly linked to the history of social movements.[4]

Solidarity and Modern Society

UN General Secretary António Guterres underlined the need for global solidarity in his opening speech at the Climate Summit in 2022: from his point of view, climate solidarity was one of the central global value concepts—as important as tolerance or freedom. Disregarding it was equivalent to global doom.[5] The more specific the UN climate negotiations in Sharm el-Sheikh became, and the closer the moment of binding contractual agreements approached, the more strongly solidarity was invoked. At the same time, however, the contours of the concept became increasingly unclear. The regulative idea of solidarity, particularly if intended for the global level, often results in generating its own disappointment and raises expectations that it can hardly fulfill. That is what the idea of solidarity has in common with other normative concepts that are highly valued by modern societies, such as justice or freedom.

The fact that climate has also been associated with the idea of solidarity may seem logical. Fundamental questions of crisis management are currently being ignited by disputes about the "right" kind of solidarity. These debates concern the social, economic, and ecological costs of current global distortions, and solidarity debates are always an expression of conflicts about the distribution of social resources in a capitalist world. Moreover, solidarity as a welfare state norm, as a semantic code, and as social practice—all these different dimensions of the term become blurred. It has been one of the "modern" values of the international labor movement since the end of the 19th century and describes a core element of the global language with respect to Catholic social teaching.[6] At the same time, solidarity has always been a political-normative concept of struggle, an instrument of power, and an ideological construct. For almost two centuries, religious and secular interpretations of modern societies have overlapped therein—with the welfare state as a specific administrative arrangement of institutionalized solidarity. In line with this tradition of discourse, solidarity may be described as a reciprocal social mode of relationship that "places being interdependent at the centre of action."[7]

But what does this mean exactly? For the early labor movement in the 19th century the reference to solidarity was an expression of the acute experience of exploitation under capitalism. This term was used to distinguish itself from older ideas of Christian "goodness" and "mercy." For Wilhelm Liebknecht, one of the leaders of the early German labor movement, solidarity had even moved to the center of a new left morality: "The concept of universal human solidarity is the highest concept of culture and morality; to realize it completely, that is the task of socialism."[8]

However for Kurt Eisner, the social-democratic Munich revolutionary and Bavaria's first prime minister after the First World War, solidarity was not only a concept of virtue but also the "builder of a quite sublime world order." The concept comprised "more than humiliating pity, more than elevating love." Solidarity, he put it in 1908, was a "cold[er], steel[ier]" word, born in the "furnace of scientific thought." In addition, it "has its cradle in the head of mankind, not in feeling. Science has nurtured it, and in the great city, among soldering and streetcars, it has gone to school. It has not yet completed its apprenticeship. But when it has matured and is all-powerful, then you will realize that in this hard concept a hot heart beats passionately in a new world of feelings."[9] At this point, Kurt Eisner had already described some of the ambivalences inherent in the history and social practice of solidarity: the longing for class-specific binding forces that arise naturally, the utopian power of a social order that transcends national boundaries and develops the extent of its commitment from the experience of social conflicts.

Debates about solidarity repeatedly touch on key areas of conflict in modern Western societies,[10] such as the relationship between political and social solidarity, between solidarity as a political concept of struggle and solidarity as a mode of social integration; the process of change and normative foundations of institutionalized solidarity, as reflected in modern welfare states, their justifications and challenges; the question of whether solidarity is to be justified rather universally or in a particularistic manner, i.e., how strongly the idea thereof is linked to family, nation, class, gender, or environmental relations; finally, the relationship between giving and receiving, the reciprocal effectiveness of solidary action and the (im)possibility of feeling connected to the "foreign" and the "other." Recently, the debate on this subject has focused on the question relating to how forms of "unconditional solidarity"[11] may evolve, that is, forms of solidarity that are not precisely based on mutual experiences or a common origin, taking notions thereof to a global—or even planetary—scale. In this context, philosopher Rahel Jaeggi has defined solidarity as a form of symmetrical-reciprocal interdependence with a non-instrumental character, the bases of which are common goals and problem perceptions. Above all, she points out that solidarity is not simply a social mode of action that has to be called upon. In her view—thus making it accessible to historians as well—it is a specific social practice that is "in some sense an artificial bond between individuals," "which is not (pre-) given, but being created."[12]

Practices of solidarity in general, and practices of climate solidarity in particular, thus do not arise from themselves but are part of purposefully created alliances. Leaving aside a more in-depth discussion on the various conceptual-historical shifts,[13] it is still important to point out that this modern concept of solidarity reflects a notion that there exists a normative bond and an obligation between people and groups to help each other, albeit a help that has always been particularized at the same time. Émile Durkheim, one of the founding fathers of sociology, described the transformation process of a mechanical, "traditional" solidarity into an "organic" one as a result of the emergence of capitalist societies in the 18th and 19th centuries based on the division of labor.[14] Solidarity hence became an integrating mechanism of modern societies that moved beyond the rationale of family ties.

With Franz Xaver Kaufmann, four elements of solidarity can be distinguished[15]: loyalty, altruism, reciprocity, and collective mutuality. While the concept of loyalty focuses on the "recognition of existing orders and the fulfillment of rules and legitimate expectations of third parties," altruism aims more strongly at the voluntariness of cooperative behavior and refers to social actions that do not require or presuppose a reciprocal response from the addressee. The concept of "extended reciprocity," as introduced by Kaufmann, refers to a social action whose motive is based on the perception of direct or indirect interdependence and interprets gifts and help as part of a specific pattern of relationships that has meaning not only in traditional societies but also in modern ones. "Collectivity-oriented behavior," as the fourth dimension, is a particularly far-reaching form of solidarity related to a larger group, namely, the solidarity of international labor or a trade union movement, for example, in which individual and collective interests overlap and whose normative rationale is not merely legal obligations but a higher, "more general" goal. Their activities go beyond those actions that, as grounded in the concept of loyalty, do only what is expected. They also suggest that self-interest and collective interest are, in part, at least congruent. The discussion on climate solidarity points to this field of tension between individual and collective expectations that transcend nation-state borders and can form new political coalitions in international climate policy.

Solidarity is not a term confined to the political left and social democracies, though. Rather, it has always been an ambiguous political fighting term, whose semantic network is by no means limited to a noble internationalism but has meanwhile also become part of the language used by the political right. In this respect, it is understood as a modern type of control regarding social action, which unfolds its effect especially within societies that are characterized less by tradition than by functional differentiation. Solidarity-based action differs from all other forms of social action by emphasizing mainly the "subordination of self-interest."[16]

A decisive factor in the use of the term is not primarily the motivation but the characteristics of social action itself. In this respect, acting in solidarity is by no means considered anachronistic, but it does enable complex societies "to experience a new kind of motivational prerequisite[17] as to individual needs for recognition and identity-forming feelings of self-esteem," the basis of which being the interpretation of a joint "community of thought" and an exceptional state of needs. A history of solidarity refers to the conflict history of capitalist society and the international labor movement, to the broad field of international aid and humanitarianism. Moreover, it reflects the history of modern welfare states and leads into the environment of new social movements and their pluralized solidarity practices since the 1970s, where the relationship with progress, growth, and structural change has been argued over. As with any good history, it never follows a straight, teleological line of development but teases out the contentions—one could say, ambivalences—of the term and its usage, which today all draw on one, or even multiple, of these historical vantage points.

Ecological-Social Transformation Conflicts

In its early days, the idea of climate solidarity was taken up primarily by those who remained suspicious of neoliberal approaches to a global climate protection policy and who often saw themselves connected to the former social movements, above all the trade union movement. For a long time the labor and trade union movements had a hard time dealing with the new organizations associated with the climate protection movement. Time and again, a central conflict of solidarity was at stake: what about job security compared to the follow-up costs of a socioecological transformation?[18]

In post-war Germany, the trade unions in the chemical, coal, and steel industries in particular, but also the automotive industry, had been shaped by this dispute over how to cope with structural change and the different growth logics. For a long time, a largely unrestrained productivity-oriented history of progress dominated within the Social Democratic Party family, which had hardly been questioned following the caesura of the global oil price shock in 1973 and the incipient debates about the ecological risks of modern industries. The fact that within the new social movements, particularly the environmental and peace movement with its many different currents, there were now quite different debates about growth and productivity, which were initially regarded by most older members of the Social Democratic Party (SPD) as the gimmicks of long-haired bourgeois sons and daughters who, ultimately, were simply doing too well.

While there were exceptions to this as well, it had been precisely the different promises for the future and such a different understanding of progress that initially put a strain on the relationship between the Social Democrats and the new social movements, among them prominently the newly founded Green Party, with Petra Kelly as one of the most prominent figureheads. In 1981, when the economic engine of the Federal Republic of Germany was sputtering and the SPD-led government was fighting against rising unemployment, one of the most prominent social democratic intellectuals of his time, the Berlin political scientist Richard Löwenthal, sparked a heated debate about the future of the party: Löwenthal, who had fled from the National Socialists to England after 1939, lamented a serious "identity crisis" in his party. His conclusion was that the Social Democratic Party was losing young voters to the so-called "alternatives," the Green party, which, this early on, was more successful in Germany than in any other county. Founded in 1980, the Greens soon enjoyed political success on the local level. In 1983, they were even voted into the Bundestag with 5.7% and 27 representatives, while at the same time, the SPD was losing more and more support among its core voters, among skilled workers, and among the residents of the cities, so that many of its own people no longer went to the polls or instead switched to the Christian Democratic Union (CDU). So, what ought to be done?[19]

Löwenthal's contribution was a direct response to a speech made by former German Chancellor Willy Brandt, who, still head of the SPD, had campaigned to win these young, "alternative" voters for social democracy; after all, Brandt said, the SPD had never been a "pure" workers' party and had always attracted other professional groups. Löwenthal was much more reserved. For unlike the young left-wing members of the student movement, who had, to the dismay of more traditionally minded Social Democrats, found their way into the SPD, he saw a categorical difference between the "drifting youths" and the "mass of the previous core voters."[20] According to Löwenthal, these "dropouts" of the "green wave" wanted to create "an alternative to *the* industrial society based on the division of labour."[21] Precisely, it was no longer about changing the existing conditions but about a radical exit:

> A party that shows understanding for the problems and the dropouts can certainly integrate some of them here and there, but only if it resolutely opposes their worldview with clear arguments. A party that avoids taking a clear stand in this confrontation can only disintegrate itself.[22]

With those who rejected the "modern world," there could be "no compromise"[23] for the SPD.

Löwenthal's diagnosis revealed at least as much about the author as about the topic itself, because in every line, one could literally sense the massive reservations toward the young voters, who were by no means conformists and who quite obviously wanted to lay the axe to the "industrial society" that was sacred to the Social Democrats. Those who wanted to create a humane world of care needed a stronger industrial society based on the division of labor, Löwenthal said—and nothing less. And the SPD—meaning Willy Brandt in particular—must on no account fall into the fallacy that these "reactionary utopias" from the margins of society had anything to do with participation.

In essence, at any rate, Löwenthal saw a threat to the broad alliance of the middle class and the working class that would have led to the SPD's success both through and after the *Godesberg Programm*, in which the SPD made its peace with the new social market economy. Löwenthal's paper found quite a few supporters, especially among the "canal workers," the unionized, and culturally rather conservative Social Democrats, who indeed had a hard time with the fact that these young people wore long hair and no ties and protested against NATO, nuclear weapons armament, for universal peace and environmental protection. But there was opposition as well, namely, on the part of Peter Glotz and Erhard Eppler, who considered Löwenthal's analysis far too crude. In order to win majorities, they said it was essential to talk to and be open with these groups and their very issues, which found sympathy among many people, especially beyond the alternative milieu, including Social Democrats. One did not have to be a squatter to realize that vacant houses and housing speculation were morally reprehensible. And anyone who was concerned about the natural foundations of life was far from being an "eco-wacko."

Indeed, the controversy over the Löwenthal paper touched on a fundamental problem of social democratic identity, the effects of which reach into the present and into debates on "climate solidarity," i.e., in view of socio-structural and socio-moral transformations, where exactly should there be any room for life strategies that define "quality of life" differently from those that are primary productivity-oriented ones? For social democracy—not only in Germany—this also remained a central field of tension because it reflected the transformation of capitalist societies in themselves, as well as their changing perceptions of risk and demands. Amid the political disputes of a maturing Federal Republic, the struggle for the future had become an important semantic battleground on which the Social Democrats felt increasingly uncomfortable.

There was still no talk of climate solidarity in the 1980s, but the debates between social democracy and green movements point to a central issue of conflict that was to shape later discussions, including international ones: could there be such a thing as "green progress"? Was this kind of progress necessary at all, given the ecological destruction, the risks of nuclear energy, and the unequal distribution of resources? For the political left in Germany, as in other European countries, above all in France and Great Britain, the debate on nuclear energy was one of the breaking points for their relationship with "socioecological" modernization, as the use of nuclear energy was linked to a specific qualitative growth model whose origins were still entirely based on the Fordist production models of the 1960s.

Introducing the "ecological modernization" formula and making it a political trademark of Oskar Lafontaine, the charismatic SPD leader between 1995 and 1998, seemed to lead to an answer to this conflict in the late 1980s that reconciled the different camps and promised a resolution of the apparent contradiction between labor and the environment.[24] For many Social Democrats, environmental protection, however sensible it may be, had ultimately been a second-order problem in view of impending mass unemployment. The new forces that wanted to be understood less as a strictly organized party and more as a "movement" appeared too unpredictable to the disciplined Social Democrats. This often had something to do with the fact that the formulaic "socioecological modernization" concealed unresolved conflicts of goals and ideas of progress, the resolution of which caused considerable friction.

Of course, the debate on environmental degradation and the climate was not without its consequences for social democracy, especially since numerous initiatives had been formed over the years, in particular with the social democratic base, who also saw themselves as part of the citizens' and environmental movement. The Social Democrats were happy to refer to Willy Brandt's famous call for "blue skies over the Ruhr" (1961) and the desire for a new "quality of life," which included better air just as much as a clean forest. The fact that Brandt's demand did not yet have much to do with environmental policy and was rather more in line with the classical social democratic tradition of labor protection did not play a particular role in this respect but sufficed for a reference needed at that point that should help secure the SPD's identity as an ecologically aware political party.

In the meantime, it had been forgotten how hard the ecological question was fought out within social democracy as early as the 1980s and how seriously these discussions were being conducted even then about a new, less technocratic understanding of progress. There was no talk about "climate solidarity" back then. With Oskar Lafontaine, however, this policy had a face as well as a voice during the years prior to German reunification. For him, a new left-wing consciousness of progress meant a socioecological reform policy based on "an ethos of ecological self-restraint,"[25] as well as the "principle of responsibility" (Hans Jonas) for market-based societies. His call for a "democratization of responsibility"[26] aimed at including more participatory elements in the decision-making process on large-scale technological projects, in order to open up the political left, again more strongly to social emancipation movements "from below," and thus also to save the "project of modernity." As passionate as these debates were about the viability of an idea of progress that was also aware of the self-destructive forces of its capitalist dynamics, and as much as they influenced the search for a new programmatic orientation at the end of the 1980s, these controversies were quickly forgotten again following the fall of the Berlin Wall and German Reunification in 1990.

The SPD's "Berlin Program" (1989) embodied this encounter history between old and new social movements, holding many different promises for the future and diagnoses of the present. With great seriousness, the Social Democrats wrestled with what they called a "future of progress," with the legacy of the Enlightenment, with a renewed distribution of social risks, and with a role of civic self-support beyond state bureaucracies. It is hard to imagine, but there were SPD-affiliated intellectuals seated with experienced members of parliament discussing Kant, Descartes, and Bacon, the contradictions of technical innovations, and the broad lines of Enlightenment history. The future was no longer purely luminous, but it was also—quite deliberately—not supported by apocalyptic visions of doom, as was believed to be the case with the alternatives and which seemed to be a key distinguishing feature of the new green milieu. Nonetheless, the social democratic attitude toward progress was no longer unbroken, either. In respect thereof, too many had read the works of Ulrich Beck and his "risk society" intensively. During the course of the 1990s, economic problems following reunification ensured that considerations about a socioecological reform policy quickly receded into the background of inner-party priorities. Whenever there had been talk of innovation and equity since the middle of the decade, it had always been about a policy of qualitative growth, where social democratic environmental politicians had been able to contribute a good idea or two, but never in the front row.

Since then, the target horizon of social democratic plans for the future has shifted significantly. This is particularly evident in the case of Olaf Scholz himself. Scholz was First Mayor of Hamburg before he became the German Chancellor, and it was during this time that he published his first comprehensive programmatic catalogue of ideas titled *Hoffnungsland* (Country of Hope). The book describes Scholz's view of the world, the consequences of globalization, migration, global political conflicts, and socio-political challenges. *Hoffnungsland* is a conceivably unpathetic text, almost as if Scholz wanted to attenuate some of the strength of its title by using as little linguistic coloring as possible. *Hoffnungsland*[27] closes with seven ideas for a just order: preserving pension insurance, strengthening the welfare state and defending the minimum wage, promoting education, affordable housing, an improved infrastructure, and a "fair tax system." His proposals for Germany's "Country of hope," environmental, and climate issues proved to be a major blank space. The same applies to his "Plea for a Society of Respect," which sought to profile the SPD as the party of labor rather than a better Green Party. For some, this was certainly not far-reaching enough, seemingly too rooted in the present, too little focused on the challenges of human-made climate change. And yet this was an interpretation of social transformation that, linguistically, conceptionally, and in terms of horizons of thought, was part of a far older tradition of social democratic search for meaning since the 1970s, and which was ultimately not about "work *and the* environment," but about "work *through the* environment."[28]

Solidarity between Work and Consumption

Current debates about the notion of climate solidarity are directly linked to the question of the "ecological consumer"[29] and the assumption that a new form of global interdependence has emerged with global warming.[30] This is not wrong. And yet the close connection between solidarity, consumption, and protest has a much longer history and different origins. Émile Durkheim had already pointed out the connection between capitalism, the modern division of labor, and the new kinds of solidarity relationships that resulted from the growing dependencies. Such interdependencies increased again significantly in the second phase of globalization after World War II and reached a completely new dimension at the end of the 20th century—and with them also conflicts about the moral quality of markets and consumer behavior. On the one hand, the extension of production chains did indeed distance consumers ever further from producers and their living conditions. On the other hand, news about poverty and injustice now reached domestic living rooms ever more quickly. Globalization made the world bigger and smaller in equal measure. Consumer criticism was not a phenomenon of the 20th century. And yet, since the 1960s, they had experienced a comprehensive expansion that went far beyond the protest forms of the labor movement and encompassed above all civil society initiatives and religious groups.

When German psychoanalyst and social philosopher Horst Eberhard Richter published his book “Learning Goal Solidarity”[31] in 1974, the hearts of a broad public went out to him. His book was to quickly become a classic of a psychoanalytic diagnosis of the times that combined a critique of capitalism and individual alienation, the history of psychic destruction with the coldness of the dominant mode of production—a story about people on the margins of society, about the critique of welfare state social bureaucracies, about the compulsion to perform, the lack of empathy, and feelings of guilt; in other words, about what needed to be exposed and overcome in order to create the conditions for acting in solidarity. Richter’s book, which sold in the tens of thousands, obviously struck a nerve that linked the debates about political solidarity with the “Third World movement” and the psycho-boom of the 1970s, which gave it its very own, as it were, “liberating” thrust. The debates about the concept of solidarity mixed very different approaches that touched on very fundamental questions about the history of solidarity: How have notions changed on this type of social relationship based on shared closeness, common interests, and emotional connection? What is the basis for the idea of mutual aid in the case of emergency and the notion of “being dependent on each other”?

The search for an “alternative economy”—similarly to more recent debates on “climate solidarity”—aimed at changing the capitalist rules of the game. In the mid-1970s, this did not go far enough for some dyed-in-the-wool Marxists, as this new “alternative economy” seemed to be a retreat into the private sphere that did not question the system. Nevertheless, the newly founded housing cooperatives and businesses sought to thwart the capitalist laws of growth and consumption.

Admittedly, what exactly the “alternative” accent would be, and how far the critique should go, remained controversial within the young alternative movement. What these new forms had in common, though, was the attempt to break through the commodity character of labor and thus the capitalist mode of production.[32] A new meaning was to be given to the various activities by producing holistically and breaking through prevailing forms of the class- and gender-specific division of labor.

In addition, the aim was to develop a form of economic activity that was more strongly oriented toward regional economic cycles and made sparing use of natural resources. In the newly founded agricultural cooperatives, self-managed farms, and businesses, it was not least a matter of rethinking the concept of alienation, which also included addressing the ecological consequences of the existing economic order. Climate issues were not yet part of this milieu, but the criticism of Western consumer societies, as formulated by the Club of Rome as early as 1972, and the assumption that there must be “limits to growth” in order to protect the ecological balance of the planet were so. The current debates on “climate solidarity” have one of their origins in these early alternative experiments.

The critique was aimed at the consequences of Fordist mass production, the industrial order that had prevailed in capitalist societies since World War I and which was entirely based on standardized mass production, assembly line production, and consumption by (pacified) workers.[33] The new, "domination-free" organizations were integrated into the broad spectrum of counter-cultural ways of life and served to create the necessary material resources of the new social movements, to produce movement-owned goods and services, and to provide daily proof that a "new life" was possible in the here and now.[34] Various debates converged on this point: the former cooperative experiments of the old workers' movement experienced a renaissance, the dilemma of which also applied to the new business start-ups, farms, cooperatives, and neighborhood newspapers. How was one to fight capitalism by means of production cooperatives, which after all carried the seeds of self-exploitation and alienation and, due to the trend toward profit orientation, either developed into exploitative enterprises themselves or dissolved?

The struggle for the sovereignty of interpretation with respect to alternative production began, as it were, at the moment of its founding.[35] Even the term "alternative economy" seemed too narrow to some because it was limited primarily to "subcultural forms of work within the alternative scene" and thus excluded, for example, "female reproductive work" or even forms of neighborly solidarity.[36] "Counter-economy" sounded too empathetic to some, and "self-economy" too egocentric.[37] It thus seemed more appropriate to speak of an autonomous or an informal sector in order to encompass the various activities in self-managed print shops as well as involvement in citizens' initiatives, churches, and associations. The central difference to traditional gainful employment, but also to the state sector, was that the goods and services produced did not follow the principle of profit maximization or symbolic gratification but were oriented to needs—and thus were part of a culture of use value. The work was particularly intensive, and it served to create its own "needs-based supply structure" that left room for self-realization and self-responsible action.[38] These new forms of production were seen as a response to a comprehensive social and economic "crisis."[39]

Mass unemployment, the consequences of rationalization, recession, the division of the workforce into privileged and marginalized groups, the "segregation of the elderly" and the lack of opportunities for the young and the migrants—all of them, together with a comprehensive state and market failure, were seen as an indication for the end of the growth-based Fordist economy with its high degree of centralization and the declining efficiency of public services. In addition to the critique of production based on the division of labor, the alternative debates since the late 1970s were primarily concerned with the demand for an abolition of the Fordist separation of work and non-work. In these debates, work and life were no longer separate spheres; instead, the representatives of various forms of alternative economies attempted to rethink the workplace as a social place, as a center of life and communication.

What is remarkable about this attempt to describe alternative economies is not the contradiction between theory and practice, which quickly wears itself out in everyday work, but the different meanings of the concepts of work, ecology, and solidarity that had taken root in parts of the alternative subculture of Western Germany, namely, the search for a pre-capitalist organization of production, the hope for the solidary character of the new workplace, the critique of consumption and the exploitation of the "Third World," and the revaluation of physical labor. Together, they seemed to be key to living "authentically," as it were, and to escaping growing alienation. Ideas about how the relationship between work and solidarity could be redefined in the future of a different industrial society had become a central subject of social conflict during the 1970s and 1980s, accompanied by fierce controversies about the reduction in working hours and the distribution of increasingly scarce social resources.

This distribution struggle also included the question of what role consumers themselves should play in the future within established or alternative production and commodity chains. For a long time, especially in the Federal Republic of Germany, mainly church groups discussed the relationship between consumption and solidarity.[40] There had already been long-standing traditions of specifically Christian support organizations for help, such as "Misereor" or "Brot für die Welt" (Bread for the world), that reached back to the late 1950s and were still partly interwoven with the activities of various mission networks, who had existed since the 19th century. But since the 1960s, the discussion has changed.[41] This was triggered by passionate debates over decolonization and the Algerian and Vietnam wars—a controversy that not only mobilized left-wing political student groups but also met with a particularly strong response from denominationally organized students. In the inner-church debates, the idea of a "word church" was now more and more frequently on the agenda and thus no longer the missionary work of lost souls but rather the option for the poorer and weakest. It was also a new theology that criticized the older idea of charity as paternalistic and instead focused more on the idea of solidarity, a comprehensive responsibility for global inequalities. This had very immediate consequences.

In West Germany, for example, it was Protestant women's groups that provided the impetus for the longest-lasting boycott action against South African goods in the late 1970s.[42] The 1979 campaign "Don't buy the fruits of apartheid" had its roots in these debates,[43] and it continued to stoke controversy about the relationship between solidarity and ethically appropriate consumer choice that had been going on for some time within the Protestant churches, especially around the World Council of Churches, in which the place of worship from the global South now also played a central role.

The question of the relationship with South Africa had been debated within Protestant churches for some time. This was especially true following the decision made by the World Council of Churches (WCC) to establish a solidarity fund as part of a comprehensive anti-racism program. The Solidarity Fund and the demands of the WCC meant active confrontation with the apartheid state, thereby touching on one of the most sensitive questions of the solidarity movements in general: What should constitute a modern, Christian sign of solidarity? Did solidarity with the oppressed even justify the use of violence in the end? There were no simple answers to these conundrums, and yet it seemed obvious that in a modern world, the question of solidarity being part of Christian values began to arise anew.

The politicization of purchasing decisions was not a new phenomenon and could be traced back to pre-modern societies; what was new, however, was the broad social impact, which initially centered on church milieus and the radical criticism of individual behavioral strategies. The various initiatives that were founded in the Netherlands, Belgium, the Federal Republic of Germany, and Great Britain faced similar challenges. Criticism of the capitalist logic of utilization was one thing, but the question was still open to what extent one should not rely on fairer trade, i.e., on the market and the decision of the consumer. Solidarity was seen as an expression of a particular way of life, as a response to post-colonial exploitation, as practiced by their own church organizations and state development aid ministries. In political practice, this expansion of the concept of solidarity led to considerable conflicts, for it was not only within the domestic churches that it remained disputed whether solidarity should not also know boundaries and whether boycotts, for example, would also harm domestic industry and consumers.

The German trade unions, for example, supported the political campaigns calling for a boycott of South Africa and emphasized international solidarity, but at the same time they vehemently advocated for closing the European textile market to Asian products. It was therefore also disputed whether the solidarity appeals did not hide new forms of paternalism and conceal protectionism.

Pluralization of the concept of solidarity since the 1970s thus not only included its excessive use within the new social movement and within the Third World movement, but it also its spread within church milieus, where the old language of mercy was increasingly—under the influence of voices from the Global South—on the defensive. Solidarity debates and the development of alternative church commodity chains for the distribution of, for example, coffee, tea, or other goods touched on central conflicts about the meaning of capitalist production. Suddenly, the decolonization movement was on the doorstep and had become part of breakfast and dinner.

Climate Solidarity and New Alliances

The concept of solidarity refers to a specific practice of social action. In debates about the emergence of "fair trade" and the consumer boycott of the 1970s, one question was already always present: How far did the change in individual behavior have to go? Didn't shopping in the "Third World store" only mean concealing exploitative conditions? Didn't such campaigns ultimately stabilize the very capitalist relations they actually wanted to change? And did the purchase of consumer goods only mean concealing a bad conscience? The historicization of such debates and moral concepts points to the contradictions that are part of such a history of solidarity, including climate solidarity. Current debates around the idea of climate solidarity and an "ecological consumer" reflect these different historical experiences of conflicts about production and consumption. They also include a new awareness of global responsibility that has emerged with the founding of the "Third World Movement." But there is also open conflict over the scope and validity of particular and universal norms of solidarity, which highlights the contradictory nature of ecological transformation as a whole. And whether climate solidarity will develop the power at a global level that UN Secretary General António Guterres has invoked is currently questionable, to say the least. So far, there is not much to suppose that it will do so.

Notes

1 United Nations, "'Cooperate or perish': At COP27 UN chief calls for Climate Solidarity Pact, urges tax on oil companies to finance loss and damage," *UN News*, November 7, 2022, https://news.un.org/en/story/2022/11/1130247.

2 United Nations, "International Human Solidarity Day 20 December," accessed May 29, 2024, www.un.org/en/observances/human-solidarity-day.

3 Giacomo Bazzani, "Climate Solidarity: A Framework and Research Agenda for Low-Carbon Behavior," *Sociological Forum* 38, no. 2 (2023): 352–374, 357.

4 Sebastian Garbe, *Weaving Solidarity: Decolonial Perspectives on Transnational Advocacy of and with the Mapuche* (Bielefeld: Transcript, 2022), 29–67.

5 Dietmar Süß and Cornelius Torp, *Solidarität: Vom 19. Jahrhundert bis zur Corona-Krise* (Bonn: Dietz, 2021); Steinar Stjerno, *Solidarity in Europe: The History of an Idea* (Cambridge: Cambridge University Press, 2004); Kurt Bayertz, "Four Uses of 'Solidarity'," in *Solidarity*, ed. Kurt Bayertz (Dordrecht: Kluwer, 1999), 3–28; Hauke Brunkhorst, "Globalizing Solidarity: The Destiny of Democratic Solidarity in the Times of Global Capitalism, Global Religion, and the Global Public," *Journal of Social Philosophy* 38, no. 1 (2007): 93–111; Arto Laitinen and Anne Birgitta Pessi, eds., *Solidarity: Theory and Practice* (Lanham: Lexington Books, 2015).

6 Michael Reder, "Solidarische Praktiken in globaler Perspektive: Sozialphilosophische Anmerkungen zu einem umstrittenen Begriff," *Widerspruch* 62 (2016): 13–26.

7 Stephan Lessenich, Michael Reder and Dietmar Süß, "Solidaritäts-Kämpfe in Zeiten der Covid-19 Pandemie," *Aviso. Magazin für Kunst und Wissenschaft in Bayern* 3 (2020): 31–33 (31).

8 Wilhelm Liebknecht, *Zu Trutz und Schutz. Festrede gehalten zum Stiftungsfest des Crimmitschauer Volksvereins am 22. Oktober 1871* (Leipzig, 1874), 14.

9 Kurt Eisner, "Sieben Briefe. An eine Freundin, IV. Solidarität," in *Gesammelte Schriften*, vol. 2, ed. Kurt Eisner (Berlin: Cassirer, 1919), 52–56 (56).

10 Stephan Lessenich, Michael Reder and Dietmar Süß, "Zwischen sozialem Zusammenhalt und politischer Praxis: Die vielen Gesichter der Solidarität," *WSI-Mitteilungen* 73 (2020): 319–326.

11 Lea Susemichel and Jens Kastner, "Unbedingte Solidarität," in *Unbedingte Solidarität*, ed. Lea Susemichel and Jens Kastner (Münster: Unrast, 2021), 13–48.

12 Rahel Jaeggi, "Solidarität und Gleichgültigkeit," in *Unbedingte Solidarität*, ed. Lea Susemichel and Jens Kastner (Münster: Unrast, 2021), 49–66 (53).

13 Michael Reder, "Solidarische Praktiken in globaler Perspektive".

14 Émile Durkheim, *Über soziale Arbeitsteilung: Studie über die Organisation höherer Gesellschaften* (Frankfurt/M.: Suhrkamp, 1992).

15 Franz-Xaver Kaufmann, "Sozialpolitik zwischen Gemeinwohl und Solidarität," in *Gemeinwohl und Gemeinsinn: Rhetoriken und Perspektiven sozial-moralischer Orientierung*, ed. Herfried Münkler and Karsten Fischer (Berlin: Akademie-Verlag, 2002), 19–34, 41–44.

16 Durkheim, *Über soziale Arbeitsteilung*, 41.

17 Ibid., 46.
18 Nora Räthzel and David Uzzel, "Trade Unions and Climate Change: The Jobs versus Environment Dilemma," *Global Environmental Change* 21, no. 4 (2011): 1215–1223.
19 Richard Löwenthal, "Identität und Zukunft der SPD," *Die Neue Gesellschaft*, December 1981, 1085–1089; quoted in Richard Löwenthal, "Identität und Zukunft der SPD," in *Soziale Demokratie: Wege und Ziele*, ed. Thomas Meyer (Bonn: Dietz, 2021), 137–145.
20 Löwenthal, "Identität und Zukunft der SPD," 138.
21 Ibid., 139.
22 Ibid., 141.
23 Ibid., 140.
24 Felix Lieb, "Von der 'ökologischen Modernisierung' zum 'Primat der Ökonomie'. Sozialdemokratische Umweltpolitik zwischen ökologischer und wirtschaftlicher Krise in den 1980er- und 1990er-Jahren," in *Parteien in der 'Krise.' Wandel der Parteiendemokratie in den 1980er- und 1990er-Jahren*, ed. Thorsten Holzhauser and Felix Lieb (Berlin: Metropol, 2021), 84–96.
25 Oskar Lafontaine, *Die Gesellschaft der Zukunft: Reformpolitik in einer veränderten Welt* (Hamburg: Hoffmann/Campe, 1988), 266.
26 Oskar Lafontaine, *Die Gesellschaft der Zukunft,* 46.
27 Olaf Scholz, *Hoffnungsland: Eine neue deutsche Wirklichkeit* (Hamburg: Hoffmann/Campe, 2017), 205–211.
28 Felix Lieb, *Arbeit und Umwelt? Die Umwelt- und Energiepolitik der SPD zwischen Ökologie und Ökonomie 1969–1998* (Berlin/Boston: de Gruyter, 2022).
29 Chris Hilson, "EU Environmental Solidarity and the Ecological Consumer: Towards a Republican Citizenship," in *Promoting Solidarity in the European Union*, ed. Malcolm Ross and Yuri Borgmann-Prebil (Oxford: Oxford University Press, 2010), 136–150; Richard Wilk, "Consumption, Human Needs, and Global Environmental Change," *Global Environmental Change* 12, no. 1 (2002): 5–13.
30 Bazzani, "Climate Solidarity," 365.
31 Horst-Eberhard Richter, *Lernziel Solidarität* (Reinbek bei Hamburg: Rowohlt, 1974).
32 Arndt Neumann, *Kleine geile Firmen. Alternativprojekte zwischen Revolte und Management* (Hamburg: Nautilus, 2008); Sven Reichardt, *Authentizität und Gemeinschaft: Linksalternatives Leben in den siebziger und frühen achtziger Jahren* (Frankfurt/M.: Suhrkamp, 2014), 319–350.
33 Rüdiger Hachtmann and Adelheid von Saldern, "Das fordistische Jahrhundert: Eine Einleitung," *Zeithistorische Forschungen/Studies in Contemporary History* 6 (2009): 174–185.
34 Sven Reichardt and Detlef Siegfried, eds., *Das Alternative Milieu: Antibürgerlicher Lebensstil und linke Politik in der Bundesrepublik Deutschland und Europa 1968–1983* (Göttingen: Wallstein, 2010).
35 Rolf Schwendter, "Notate zur neuesten Geschichte der alternativen Ökonomie," in *Die Mühen der Ebene: Grundlegungen zur alternativen Ökonomie, Teil 2*, ed. Rolf Schwendter (München: Verein zur Förderung der sozialpolitischen Arbeit, 1986), 60–79 (61ff.).
36 Rolf G. Heinze and Thomas Olk, "Eigenarbeit, Selbsthilfe, Alternativökonomie: Entwicklungstendenzen des informellen Sektors," in *Die Mühen der Ebenen: Grundlegungen zur alternativen Ökonomie, Teil 2*, ed. Rolf Schwendter (München: Verein zur Förderung der sozialpolitischen Arbeit, 1986), 113–126 (114f.).

37 Rolf Schwendter, "Notate zu den Mühen der Berge," in *Die Mühen der Berge: Grundlegungen zur alternativen Ökonomie, Teil 1*, ed. Rolf Schwendter (München: Verein zur Förderung der sozialpolitischen Arbeit, 1986), 286–292.
38 Heinze and Olk, "Eigenarbeit, Selbsthilfe, Alternativökonomie," 115.
39 Martin Geyer, *Geschichte der Sozialpolitik in Deutschland seit 1945, Bd. 6: Die Bundesrepublik 1974 bis 1982: Neue Herausforderungen, wachsende Unsicherheiten* (Baden-Baden: Nomos, 2008).
40 Peter van Dam, "Handel im Tempel? Fair Trade und Kirchen in den Niederlanden seit 1945," in *Neue Soziale Bewegungen als Herausforderung sozialkirchlichen Handelns*, ed. Wilhelm Damberg and Traugott Jähnichen (Stuttgart: Kohlhammer, 2015), 279–296.
41 Süß and Torp, *Solidarität*, 101–109.
42 Sebastian Tripp, *Fromm und politisch: Christliche Anti-Apartheid-Gruppen und die Transformation des westdeutschen Protestantismus 1970–1990* (Göttingen: Wallstein, 2019).
43 Sebastian Tripp, *Fromm und politisch*, 107–161.

3 Transformation and/of "Colonial Tropes"

Latin American Narrative Palimpsests

Victor A. Ferretti

... and the woods were their garden.

A. von Humboldt[1]

Introduction

As global health issues, the so-called Great Recession, or international tensions surrounding food and energy supply have shown, among others, crises continue to emerge worldwide in the 21st century. Nonetheless, if reflectivity is a signature of "global risk societies," as stressed by German sociologist Ulrich Beck,[2] the necessity to not only realize their ambivalences but also to re-analyze systemic one-sidedness has become a priority in contemporary transformation research and practices that seek to avoid the (re)production of economic, ecological, and societal emergencies that come with the asymmetric (or all too complexity reducing) dynamics that mark our heavily interconnected global reality.

The tensions surrounding energy transition are a case in point: in a post-industrial scenario, e-mobility seems to be a central climate-friendly alternative to transportation running on fossil fuels. At the same time, local and regional populations in countries like Argentina, Bolivia, Brazil, and Chile are facing environmental justice and pollution issues that come with the lithium industry—and which tend to be considered less pressing[3] and subordinate[4] to a pronounced "worldwide necessity" for more e-mobility and interwoven macroeconomic interests.[5] Even so, by prioritizing climate protection at the expense of environmental protection and justice, an ambivalence is made explicit[6] as a result of inequality—indicating, at the same time, contingency (i.e., the possibility of considering, for example, the protection of the environment as being not less important).[7]

In view of the fact that literary fiction, since Aristotle's *Poetics*,[8] is a classic medium for reflective worldliness and "worldmaking,"[9] and bearing in mind that the Americas have represented since Early Modern times a continental "palimpsest" full of European "exploitation marks," this chapter will focus on Latin American

DOI: 10.4324/9781032627984-5

novelas de la selva ("jungle novels"), namely a literary genre of the 20th century that is dedicated to the problematization of colonialism and the consequences of the so-called rubber boom in South America in the second half of the 19th and early 20th centuries. By citing the complexity of the Latin American *selva* (Spanish and Portuguese for "jungle"/"forest")[10] as a dispositive way to rethink—in a critical transcontinental and discourse line of tradition—certain exploitation dynamics (relating to silver, guano, rubber,[11] tannin,[12] lithium, etc.) still relevant these days, an ecocritical approach will be followed. We should not forget that the "forefather of sustainability," Hans Carl von Carlowitz, recorded in Chapter 4 (§ 1) of his *Sylvicultura Oeconomica* (1713) an atrocious European "*oíko*-culture" of natural extinction in the context of "primitive" American woods that can be considered prototypical of "*oíko*logic" ambivalence.[13]

The proposition is that an ecocritical approach to the history of the discourse of substances can be helpful in avoiding the notion that required socio-ecological transformation turns unintentionally into an "imperialistic" trope. Moreover, emerging environmental ambivalence, that is the principal argument of the following "deciphering" of the *selva*, can serve as a reflective indicator thereof.

Welcome to the *Selva*

In light of the "jungle" being a polyvalent symbol of "primitiveness," "viciousness," and "alterity,"[14] it is no coincidence that, after European jungle textures flourished "imperi(c)ally" in the long 19th century,[15] a (post-)modern series of *selva* fictions has sprouted throughout American literatures in the last century. An early example of the tradition of the *novela de la selva* is a tropical novel by an Anglo-Argentine writer named William Henry Hudson, in whose *Green Mansions* (1904) the Venezuelan rainforest functions as a projection space for Rousseau-esque *noble savageness*—[16] later impersonated by Audrey Hepburn in Mel Ferrer's motion picture (USA 1959) of the same title.

Half a century after Hudson, in Mario Vargas Llosa's *La casa verde* (*The Green House*) from 1966, the exotically attractive *selva* has become a green "brothel" where a forcibly proselytized indigenous girl from the Peruvian Amazonas has been stranded, converting the jungle into a simile[17] for civilizational *barbarie*[18] and violent cupidity. The latter had previously been raised socio-critically by Upton Sinclair's revelation novel *The Jungle* (1906), scandalizing the atrocious working conditions in the US-American food industry of the early 20th century by stressing—metaphorically—the inhuman dimension of the abovementioned "jungle."

Besides "geoliterary" variety, it is noteworthy that in all three American jungle novels, "wilderness" functions as a heterotopic trope— precisely, a "heterotrope." According to Michel Foucault's conception of *espaces autres* from 1967,[19] heterotopies (i.e., other spaces) such as brothels or meatpacking plants are less external than implicit spaces of society that might displace commercial carnal desire or industrial slaughtering to zones et al. on the outskirts: its exploitation and subjugation practices are still an integral part of "normal" society.

This is also why, in Walter Ernest Hardenburg's *Putumayo* account *The Devil's Paradise* from 1912, *the atrocities committed upon the [enslaved] Indians* in the Anglo-Peruvian Amazon Rubber Company[20] disclose the *selva* as an unequal space per se of exploited (rubber) exploitation (slavery).

A decade later, in José Eustasio Rivera's *La vorágine* (*The Vortex*; 1924), the rainforest becomes not only a greedy prison, devouring its human captives,[21] but reveals also a resilient Medusan character,[22] *defending itself from its executioners*.[23] Reading this archetypical novel a century later, it seems that the actuality of the Riverian wilderness has gained importance concerning "the human effect."[24]

Moreover, we may consider Alejo Carpentier's *Los pasos perdidos* (*The Lost Steps*; 1953), in which a sublime jungle texture translates into a kind of sanctuary for a Latino musicologist from North America. As León Hazera notes, this is the result not only of a marvelous-realistic *aisthesis* (perception) of space and time,[25] but also of a classic reflexivity that is based on the urban-rural dichotomy through which residents of the *selva* would prefer its atrocities to more "civilized" ones.[26] Nonetheless, it would be shortsighted to reduce the metapoetic *Lost Steps* to a *golden* escapist tale. Especially, as Wylie underscores,[27] Carpentier's exploration of the *selva* is revealed as a "poststructuralist" working off Western travel literature *topoi* and apperceptions (or predeterminations) of "the" jungle. Consequently, after leaving the "Edenic," palimpsestic *selva*, the central character can no longer return to it, nor can he go home *without* stating semiosis (i.e., sign suspension):

> One day [man] makes the irreparable mistake of wandering off the beaten track, believing that the exceptional can be so twice, and, on returning, he finds the landscapes disturbed, the reference points swept away, while the informants have changed their appearance…[28]

His differential *anagnorisis*, to be "unable to be once again who I was,"[29] interlinks the protagonist of *Los pasos perdidos* with one of the most ingenious fictions of the Early Modern period, namely Jacopo Sannazaro's *Arcadia* (1504), wherein a Naples urbanite flees to the fictional land of the herdspeople, Arcadia, to discover that one cannot escape one's own "fictions."[30] In the case of Carpentier's novel, the discursivity framing the perception of the *selva* as a metonymy for Latin America seems to cry out for suspension, admitting other "realities" and promoting a transformed perception of worldliness (and not only wilderness).

But perhaps no other *novela de la selva* has been as successful in evoking the jungle as a risky, ambivalent space as Mario Vargas Llosa's second novel,[31] *La casa verde* (*The Green House*). The Peruvian Nobel Prize winner (2010) and member of the Académie française (2023) assembles in this master narration several storylines orbiting around a re-erected brothel (the "Green House") in historic Piura, the lives of the converted *aguaruna* girl "Bonifacia" (and later the prostitute Selvática) and her indigenous father, and of a Japanese Brazilian contrabandi(s)t named Fushía, all of whom compose an artful structure full of colonial, capitalist, and patriarchal cultures of violence. And it is surely the female protagonist and heroine's hybrid identity that provokes a "*Green House* effect," since the anonymous indigenous girl that is baptized "Bonifacia" by missionary nuns and converted by a misogynic society into the *Casa Verde* prostitute *re*named "Selvática," is literally a hetero*tropic* self. More than *ambient*,[32] the jungle ("casa verde") turns out to be ambivalence *in natura*, as one episode, highlighted by León Hazera,[33] forcefully proves by presenting Bonifacia as an accomplice in the escape of two abducted indigenous girls from the *selva*tic mission, trying to open *the* "exit door":

> The third key turns, the door gives way, but outside there must be a tenacious concentration of stems, scrubs and climbing plants, nests, spider webs, fungi and vines that resist and block the door. Bonifacia rests her whole body on the wood and pushes—there are very light, multiple rips and a brittle rumble—until a sufficient opening is formed. She holds the door open, feels the touch of soft filaments on her face, hears the murmur of the invisible foliage, and suddenly, behind her back, another murmur.[34]

Anthropomorphically, the *selva* tries here to hinder two indigenous girls, once kidnapped from it, escaping from the jungle mission and to come "home." Yet, in the differential sense of "the" erased *selva* in Carpentier's *Lost Steps*, it is not possible for them to *re*turn. The double transgression (break out and in) in this scene is justly the expression of a *selva*tic ambivalence (here, domesticated wilderness).[35] Or, more specifically, the colonized jungle rejects the *re*turn of its missionized natives, having become a palimpsestic *selva* full of forceful inscriptions that might be eroded and transformed by newer ones, even though this cannot hinder the emergence of ambivalence as a product of controversial and intrusive practices.

The reason now why *novelas de la selva* like Vargas Llosa's *Green House* are so valuable, not only for ecoliterary but also environmental discursive history, is that—as *explicit* fictions—they make us aware of traded imperialistic inscriptions, on the one hand, and the *implicit* fictionality of their narratives, on the other. Furthermore, as in an era marked by dense and complex entanglements on a global scale, emerging challenges cannot be "solved" by neglecting the other, without evoking further emergencies.

More than half a century after they were first published, Latin American *jungle books* continue to expose these ambivalent dynamics and relations that continue to this day to influence the critical relationship between "postcolonial" societies and Occidentalism: a hereditary "colonializational risk," to put it one way, that Alexander von Humboldt (1769-1859) outlined in his posthumous *American Travel Journals*:

> They [missionaries] have caused him [native] to worry about the future, which strains his powers. They have founded a work whose consequences for this hemisphere are not calculable, a work which will be more permanent than their teaching, and [...] will bring about, in the long succession of centuries, changes that will destroy this teaching itself and these punishments.[36]

Transformative Ambivalence

As Homi Bhabha evinced in 1984 from a postcolonial perspective,[37] *ambivalence* forms the basis of "colonial discourse,"[38] generating certain subversions that can become challenging for hegemony itself.[39] Anyhow, being not only the expression of an equivocation or inner contradiction (to be resolved), but also problem-bearing one-sidedness, ambivalence can also be revealing when it comes to conflict resolution in the context of transformation and the discourse that evolves around it.

One can in fact argue that ambivalences draw our attention precisely to tendentious "problem-solving." In addition, among the various global threats with which our societies must deal, the risk of returning to imperialist discursivity[40] in the name of "change" is a paradoxical—albeit concrete—one, since the risks of global warming also comprise the risk of actualizing hegemonic practices and narratives as a root cause, thus reinscribing inequality[41] in Latin American palimpsests as already seen in Potosí, Manaus, La Gallareta, etc.[42] So, taking advantage of regional "poverty" for change's sake, legitimizing *subsequent problems*[43] with global ends seems not only anachronistic but also controversial: as the example of water shortages reveals, environmental risks are often more than "merely" a regional threat but emerge on a global scale.[44] Accordingly, saving the global climate at the expense of regional environmental issues is an ambivalent premise, and whereas transforming world risk societies without inequality might sound difficult, it nevertheless continues to be a realistic proposition.

By understanding that ambivalence is less a conflicting constellation than (already) a product of problematic unilateralization, the chances of (re)thinking the other (side, direction, position, etc.) of a complex situation and its contingencies (i.e., alternate possibilities) become more evident, as the risk of creating further risks because of monocentric decisions can come to be systematic. Explicit fictions like the *novelas de la selva*, based on historic rubber exploitation practices in the name of industrial innovation and profit, invite us *not* to forget that ideology, etc. is also fictitious, prioritizing one aspect (*as if* this would be the most important one).[45] In times of global social risks, it may be helpful to remember that unilateralism through ambivalence is an "emergence case" of complexity.

Consequently, without having to be topically *salina* novels, Latin American "jungle books" can help us to comprehend better continuities in the imperialistic tradition of what may be called the *golden* trope in South America, starting with Early Modern silver mining over the extractions of guano and rubber in the longer 19th century up to lithium exploitation in the 21st century (all four "white golds"), including *quebracho* tannin ("red gold")[46] production in the 20th century. All these transcontinental inscriptions, since Columbus et al., constitute a meta-discursive Latin American palimpsest, needing not only to be decoded, but also to be decolonized for sustainable change.

Notes

1 Alexander von Humboldt, "Die Kultur des Menschengeschlechts," in *Das Buch der Begegnungen* (Munich: Manesse, 2018), 51. The quote is my translation from the original: "…und der Wald war ihr Garten." NB: To facilitate our understanding of the argumentation of this chapter based on primary sources, all English citations from German and Spanish original texts have been translated with help of the PONS text translator (https://en.pons.com/text-translation) and then revised by me, prioritizing content over literariness.

2 See Ulrich Beck's tracking of the *Risikogesellschaft* (1986) around the Chernobyl disaster and a *Weltrisikogesellschaft* (2008) amid the global financial crisis at the beginning of the 21st century.

3 *Instat tempus* argumentation, contrary to environmental protection objections, is also part of climate disaster control politics in Central Europe, as a prominent high-water case in the Austrian province of Salzburg from summer 2021 shows (see Jan Peter, "Ist er schuld am Hochwasser von Hallein?" *Spiegel*, July 25, 2021, www.spiegel.de/ausland/unwetter-katastrophe-in-oesterreich-ist-der-naturschutzbund-schuld-am-hochwasser-von-hallein-a-82287df0-b627-4f20-8d6a-af225f791a2e.

4 See Ulrich Beck, *Risikogesellschaft* (Frankfurt/M.: Suhrkamp, 2015), 40–42.

5 In a German television interview from December 2022, the CEO of Mercedes-Benz Group, Ola Källenius, compared the environmental damage caused by lithium mining in South America to collateral water damage (!) caused by firefighting (see Ola Källenius, "Exklusiv-Interview mit Mercedes-Chef Källenius," interview by Fritz Frey and Alex Seker, *Doku & Reportage*, SWR, December 20, 2022, video, 20:37–22:56, www.ardmediathek.de/video/doku-und-reportage/exklusiv-interview-mit-mercedes-chef-kaellenius/swr/Y3JpZDovL3N3ci5kZS9hZXgvbzE3Nzg3Nzc. Regarding the legitimation of *latent side effects* ("latente Nebenwirkungen"), see Beck, *Risikogesellschaft*, 45.

6 This understanding of ambivalence as a result (and less constellation) is based on Franz Fischer's "atensionality," as commented by Thomas Altfelix (see Franz Fischer, *Proflexion und Reflexion* (Vienna: Passagen, 2007), 145–49), and critical discussions with participants at my Ibero-Romance seminar on the "Novela de la selva" in the summer semester of 2023 at the University of Augsburg.

7 Violent protests in June 2023 against a constitutional reform compromising environmental justice in Northern Argentine Jujuy (lithium) province are proof of this dynamic (see Cristina Papaleo, "Litio en Argentina: 'El lado oscuro de la electromovilidad'," *Deutsche Welle*, June 30, 2023, www.dw.com/es/explotaci%C3%B3n-de-litio-en-argentina-el-lado-oscuro-de-la-electromovilidad/a-66084159.

8 "From what we have said, it will be seen that *the poet's function is to describe, not the thing that has happened, but the kind of thing that might happen, i.e., what is possible as being probable or necessary*. The distinction between historian and poet is not in the one writing prose and the other verse […]; it consists really in this, that the one describes the thing that has been, and the other [poet] *a kind of thing that might be*. Hence poetry is something more philosophic and of graver import than history, since its statements are of the nature rather of universal, whereas those of history are singulars" [Aristotle, *Art of Poetry* (London: OUP, 1952), 25; transl. Bywater-Hamilton Fyfe].

9 Or, as Nelson Goodman (con)stated in his groundbreaking *Ways of Worldmaking* (Indiana: Hackett, 1978), 102: "The worlds of fiction, poetry, painting, music, dance, and the other arts are built largely by such nonliteral devices as metaphor, by such nondenotational means as exemplification and expression, and often by use of pictures

or sound or gestures or other symbols of nonlinguistic systems. Such worldmaking and such versions are my primary concern here; for a major thesis of this book is that the arts must be taken no less seriously than the sciences as modes of discovery, creation, and enlargement of knowledge in the broad sense of advancement of the understanding, and thus that the philosophy of art should be conceived as an integral part of metaphysics and epistemology."

10 As the young Charles Darwin noted ambivalently in September 1836, South American rainforests are naturally not limited to the Amazonian region: "Among the scenes which are deeply impressed on my mind, none exceed in sublimity the primeval forests undefaced by the hand of man; whether those of Brazil, where the powers of Life are predominant, or those of Tierra del Fuego, where Death and Decay prevail" [Charles Darwin, *Voyage of the Beagle* (London: Penguin, 1989), 374.].

11 Rubber plantations continue to be problematic today. See Nathalie Bertrams et al., "Europas Gier nach Gummi," *Süddeutsche Zeitung*, July 7, 2023, www.sueddeutsche.de/projekte/artikel/wissen/kamerun-kautschuk-rodung-lieferkettengesetz-e276333/?reduced=true. The article deals with tropical Cameroon.

12 Compare the notorious case of the London-based Forestal Land, Timber & Railways Company Ltd. that until the early 1960s exploited *quebracho* trees for tannin extraction in the Southern Chaco regions of Argentina in a neocolonial manner with an infrastructured exploitation system that oppressed workers and caused severe environmental damage [see Teresa Arijón, "La Forestal, la firma inglesa que explotó los bosques de quebracho santafesinos," *La Nación*, August 8, 2018, www.lanacion.com.ar/turismo/la-forestal-la-triste-historia-de-los-bosques-de-quebracho-santafesinos-nid2158551/; Anacarsis L. Acevedo, *Investigación a la Forestal* (Buenos Aires: CEAL, 1983)].

13 See Hans Carl von Carlowitz, *Sylvicultura Oeconomica* (Leipzig: Braun, 1713), 41–42, www.digitale-sammlungen.de/en/view/bsb10214444?page=,1. – Aforementioned (see note 13) *quebracho* tree exploitation practices in the 20th century by British Forestal Company in Argentina would be a transatlantic testimony for a long-lasting European "οἶκο-culture" of forest extinction.

14 See Victor A. Ferretti, "Dschungel," in *Metzler Lexikon literarischer Symbole,* ed. Günter Butzer and Joachim Jacob (Stuttgart: Metzler, 2021), 120–121.

15 Examples would be the Aouda episode in the Indian jungle (chap. 12–14) in Jules Verne's *Around the World in Eighty Days* (1873), Kipling's *Jungle Book* (1894/95) or Arthur Conan Doyle's *Lost* (Amazonian) *World* (1912) tour, where *empire*listic worldviews dominate.

16 For an alert reading of this novel, see Lesley Wylie, *Colonial Tropes and Postcolonial Tricks* (Liverpool: LUP, 2009), 20–23.

17 The green eyes of the—first converted and then prostituted—indigenous girl function as a *tertium comparationis* for the Amazonas and the Piura brothel named "Green House," as Lydia de León Hazera [*La novela de la selva hispanoamericana* (Bogotá: Instituto Caro y Cuervo, 1971), 228] remarks.

18 It seems worth taking into account here that the later Argentine president Domingo Faustino Sarmiento (1811–1888) promulgated in his expository *Civilización y barbarie* (1845) the 'classic' dualism between urban-civilized and rural-barbarian Latin American dynamics.

19 See Michel Foucault, "Des espaces autres," *Empan* 54, no. 2 (2004): 12–19, www.cairn.info/revue-empan-2004-2-page-12.htm.

20 Mario Vargas Llosa dedicated the Irish uncoverer Roger Casement 2010 a 'Global South novel' entitled *El sueño del celta* (*The Dream of the Celt*), who had diplomatically

denounced not only colonial Congo but also Anglo-Peruvian atrocities in the beginning of the 20th century.

21 See León Hazera, *La Novela de la selva hispanoamericana*, 136–140.

22 "Dijo que los árboles de la selva eran gigantes paralizados y que de noche platicaban y se hacían señas. Tenían deseos de escaparse con las nubes, pero la tierra los agarraba por los tobillos y les infundía la perpetua inmovilidad. Quejábanse de la mano que los hería, del hacha que los derribaba, siempre condenados a retoñar, a florecer, a gemir, a perpetuar, sin fecundarse, su especie formidable, incomprendida. El Pipa les entendió sus airadas voces, según las cuales debían ocupar barbechos, llanuras y ciudades, hasta borrar de la tierra el rastro del hombre […]" [José Eustasio Rivera, *La vorágine* (Bogotá: BNC, 2015), 154, https://web.seducoahuila.gob.mx/biblioweb/upload/La-voragine.pdf]: "He said that the trees in the jungle were paralyzed giants and that at night they talked and signaled to each other. They longed to escape with the clouds, but the earth seized them by the ankles and infused them with perpetual immobility. They complained of the hand that wounded them, of the axe that knocked them down, always condemned to sprout, to bloom, to groan, to perpetuate, without fecundating, their formidable, misunderstood species. El Pipa understood their angry voices, that they should occupy the wasteland, and the plains, and the cities, till they blot out the foot of man from the earth […]" (my transl.).

23 "La selva se defiende de sus verdugos, y al fin el hombre resulta vencido" (Rivera, *La vorágine*, 186): "The forest defends itself from its executioners, and at the end man is defeated" (my transl.).

24 See Hermann Herlinghaus, "Rivera, José Eustacio: *La vorágine,*" in *Kindlers Literatur Lexikon*, ed. Heinz Ludwig Arnold (Stuttgart: Metzler, 2020), 187, https://doi.org/10.1007/978-3-476-05728-0_17571-1.

25 "El ambiente de la selva riveriana [en *La vorágine*] persiste en la obra de Carpentier [*Los pasos perdidos*], con la diferencia de que en ésta la perspectiva mítica matiza los elementos peligrosos de la selva" (León Hazera, *La novela de la selva hispanoamericana*, 198): "The atmosphere of the Riverian jungle [in *The Vortex*] persists in Carpentier's work [*The Lost Steps*], with the difference that in this work the mythical perspective nuances the dangerous elements of the *selva*" (my transl.).

26 "Es indudable que la naturaleza que aquí nos circunda es implacable, terrible, a pesar de su belleza. Pero los que en medio de ella viven la consideran menos mala, más tratable, que los espantos y sobresaltos, las crueldades frías, las amenazas siempre renovadas, del mundo de *allá*" [as cit. in León Hazera, *La novela de la selva hispanoamericana*, 215; Alejo Carpentier, *Los pasos perdidos* (Buenos Aires: Schapire, 1975), 193]: "There is no doubt that the nature that surrounds us here is implacable, terrible, despite its beauty. But those who live in the midst of it consider it less bad, more treatable than the horrors and shocks, the cold cruelties, the ever-renewed threats of the world *beyond*" (my transl.). – Also forging a link to certain *oíko*logic colonial practices (see note 13).

27 See Wylie, *Colonial Tropes*, 29–34.

28 My translation of: "Un día [el hombre] comete el irreparable error de desandar lo andado, creyendo que lo excepcional pueda serlo dos veces, y al regresar encuentra los paisajes trastocados, los puntos de referencia barridos, en tanto que los informadores han mudado el semblante…" (Carpentier, *Los pasos perdidos*, 267).

29 My translation of "incapaz de ser otra vez el que fui" (Carpentier, *Los pasos perdidos*, 272).

30 See Victor A. Ferretti, *Bucolica mente* (Kiel: CERES, 2017), 60–63.

31 For more details. see Thomas M. Scheerer, *Mario Vargas Llosa* (Frankfurt/M.: Suhrkamp, 1991), 21–36.

32 See León Hazera, *La Novela de la selva hispanoamericana*, 218–247 (246).
33 See León Hazera, *La Novela de la selva hispanoamericana*, 229–230.
34 My translation of: "La tercera llave gira, la puerta cede, pero afuera debe haber una tenaz concentración de tallos, matorrales y plantas trepadoras, nidos, telarañas, hongos y madejas de lianas que resisten y atajan la puerta. Bonifacia apoya todo su cuerpo en la madera y empuja – hay levísimos, múltiples desgarramientos y un rumor quebradizo – hasta que se forma una abertura suficiente. Sujeta la puerta entreabierta, siente en su cara el roce de suaves filamentos, escucha el murmullo del follaje invisible y, de pronto, a su espalda, otro murmullo" [Mario Vargas Llosa, *La casa verde* (Barcelona: Círculo de Lectores, 1974), 70].
35 Consequently, the later Selvática (and converted Bonifacia) anagnoristically professes: "I became like them, Mother" (own translation of: "Me volví como ellas, Madre"; Vargas Llosa, *La Casa verde*, 70).
36 My translation of: "Sie [Missionare] haben ihm [Urbewohner] Sorgen für die Zukunft erregt, die seine Kräfte spannen. Sie haben ein Werk gegründet, dessen Folgen für diese Hemisphäre nicht zu berechnen sind, ein Werk, das dauerhafter als ihre Lehre sein, ja in der langen Folge der Jahrhunderte […] Veränderungen hervorbringen wird, welche diese Lehre selbst und diese Ahndungen vernichten wird" (Humboldt, "Die Kultur des Menschengeschlechts," 50).
37 See Homi Bhabha, "Of Mimicry and Man: The Ambivalence of Colonial Discourse," *October* 28 (1984): 125–133, https://doi.org/10.2307/778467.
38 "[…] [T]he discourse of mimicry is constructed around an *ambivalence*; in order to be effective, mimicry must continually produce its slippage, its excess, its difference. The authority of that mode of colonial discourse that I have called mimicry is therefore stricken by an indeterminacy; mimicry emerges as the representation of a difference that is itself a process of disavowal" (Bhabha, "Of Mimicry and Man," 126).
39 "The *menace* of mimicry is its double *vision* which in disclosing the ambivalence of colonial discourse also disrupts its authority" (Bhabha, "Of Mimicry and Man," 129); "The ambivalence of mimicry […] suggests that the fetishized colonial culture is potentially and strategically an insurgent counter-appeal" (Bhabha, "Of Mimicry and Man," 131).
40 Even though Edward W. Said does *not* treat—in a certain postcolonial tradition (see Fernando Coronil, "Beyond Occidentalism: Toward Nonimperial Geohistorical Categories," *Cultural Anthropology* 11, no. 1 (1996): 51–87)—Iberian *Culture and Imperialism*, his anglo- and francocentered interest on "imperialism" and "colonialism" [see Edward W. Said, *Culture and Imperialism* (London: Vintage, 1994), 8–10] can nonetheless be considered relevant for a critical Global South approach.
41 Still pertinent in this context is not only Beck's 1986 radiography of global risk industrial dynamics in *Risikogesellschaft,* 54–58, but also his later sketch of global inequality dynamics in *Weltrisikogesellschaft* (Frankfurt/M.: Suhrkamp, 2008), 285–291.
42 For the latter locality in Argentina, see note 13; for the two well-known South American colonial *topoi*, see "Potosí," *Encyclopedia Britannica*, August 7, 2015, www.britannica.com/place/Potosi-Bolivia; "Manaus," *Encyclopedia Britannica*, December 3, 2023, www.britannica.com/place/Manaus.
43 See Beck, *Risikogesellschaft*, 26–27.
44 "El agua vale más que el litio" (*Water is worth more than lithium*) is one of the slogans (photo 4) of the Salinas Grandes indigenous activists in Northern Argentina (see Felix Dorn, "'No al Litio'," *Spiegel*, June 19, 2019, photo series, www.spiegel.de/fotostrecke/lithium-foerderung-in-argentinien-streit-um-salinas-grandes-fotostrecke-169238.html.

45 See *in extenso* Hans Vaihinger, *Die Philosophie des Als Ob* (Leipzig: Meiner, 1922), 21–327.

46 See Pablo Benito, “La Forestal y el ‘oro rojo’,” *El Litoral*, June 9, 2023, www.ellitoral.com/opinion/the-forestal-land-timber-and-railways-company-limited-ruta-quebracho-colorado-oro-rojo-villa-guillermina-documental_0_aPb2MXS88Q.html.

4 Interactions of Efficiency, Consistency, and Sufficiency across Levels

Assessing Innovation Tensions through the Lens of Paradox Theory

Marcus Wagner

Introduction

In economics and management theorizing, there is a longstanding tradition arguing in favor of a business case for sustainability, which implies a win-win logic, in that businesses can support sustainability and derive economic profit from doing so.[1] Critics of this approach have argued that this win-win logic ignores a paradox between micro-level progress and macro-level environmental deterioration.[2] For instance, solar energy and electric mobility, as the main elements of a global large-scale ecological transformation, can still be connected to polluting business practices, as Hernández Westphal's chapter on lithium-silver mining in Mexico in this book vividly illustrates. While those "green" minerals are core components of renewable energy technologies that play a crucial role in mitigating greenhouse gases, they are sourced via an intensive extraction process accompanied by high environmental and social costs. Moreover, as Westphal illustrates, companies can create a positive narrative of contributing to longer term ecological transformation while creating significant additional environmental deterioration from current production growth.[3] Cautioning more generally against the win-win hypothesis is the Impact-Population-Affluence-Technology (IPAT) equation, developed by biologists Barry Commoner and Paul Ehrlich and physicist John Holdren in 1972, the year of the first United Nations Conference on the Human Environment in Stockholm. IPAT mathematically calculates human impact on the environment (I) as the multiplicative effect of the human population on earth (P), the level of affluence (A, measured as per capita income), and a technology factor (T), which captures efficiency: $I = P * A * T$.[4] The IPAT equation suggests that alongside efficiency as one key business case mechanism, sufficiency and consistency are equally relevant drivers for environmental outcomes. While efficiency is addressed in the IPAT equation by means of a technology factor, sufficiency can be gauged as the inverse of affluence, i.e., it reflects individuals' willingness to abstain or abdicate from material consumption. Mathematically, this willingness is the inverse of per capita income in the IPAT equation. Finally, consistency is conceptually a very broad term comprising the avoidance of overpopulation or resource overuse,

DOI: 10.4324/9781032627984-6

achieving high levels of circularity, and the usage of substances that already exist in nature; it is captured more narrowly in the IPAT equation by population level.[5]

The IPAT equation, originally developed by three natural scientists in the 1970s pursuing an ecological economics, rather than a mainstream economics approach, challenges the notion that efficiency alone is a viable strategy for large-scale ecological transformation. Rather, the IPAT equation suggests that other factors, such as population and affluence, need to be considered as well. The three factors are furthermore not independent but interdependent. To illustrate this point, efficiency gains can lead to increased affluence (for example, by means of rebound effects, which shall be discussed in more detail below). Equally, technological improvements and higher income levels can increase population levels due to improvements in healthcare and medical support.

Sufficiency thus needs to be a complement to efficiency in large-scale ecological transformations, and the same applies to consistency in its variety of interpretations. The IPAT equation also suggests that different factors may need to or can be traded off. Such trade-offs, which potentially create strong to insurmountable tensions, can be understood as a special case of ambivalence. A more recent theoretical development in the field of management, namely, the so-called paradox theory, supports this interpretation. As this contribution wishes to illustrate, paradox theory enables management scholars and economists (but potentially also other scientific disciplines) to address tensions and trade-offs, and in doing so ultimately helps to address the idea that corporate sustainability actions can still be complicit in creating global environmental damage. This in turn helps to answer the following question: Under what conditions can technology and innovation lead to firms maximizing contributions to support sustainable development?

In the remainder of this chapter, I explain in more detail this recent theorizing in the economics and management domain about tensions and paradoxes (both of which will be defined in the next section) and then link this with studies focusing on the role of innovation and technology for global sustainability. As part of this approach, I attempt to integrate different literatures relating to the environmental humanities in terms of efficiency, consistency, and sufficiency, as well as ecological modernization and philosophical views on technology with paradox theory. In my analysis, I develop a multi-level perspective to account for interactions between the macro-, the meso-, and the micro-levels with regard to tensions, sustainability, and technology.[6]

A Primer on Paradox Theory

The notion of paradox, initially of Greek origin and referring to a (seemingly) unavoidable contradiction, is increasingly used in management studies in the context of sustainability-related debates to refer to issues of appropriate behavior of actors such as businesses, governments, or consumers in unsuitable incentive structures, even to the degree that trade-offs become unfeasible. An illustrative example of such a paradox is when these actors become increasingly concerned and active in supporting environmental protection or social justice, but global environmental quality or social equality continue to worsen. More specifically, in the context of management research, paradoxes have been defined as "contradictory, yet interrelated, demands embedded in an organization's goals,"[7] and therefore paradoxical tensions refer to "[…] elements that seem logical in isolation but absurd and irrational when appearing simultaneously"[8]; for instance, this would be the case if a business wants to increase photovoltaic energy generation at their sites but would need to cut down many trees to do so.

Specifically, this implies that tensions become paradoxical when competing elements are simultaneously contradictory and interrelated and persist over time.[9] A paradox lens in such a situation aims to find ways to simultaneously address competing objectives in a so-called both/and thinking[10] approach (as opposed to either-or thinking of mutually exclusive alternatives or choices). Both/and thinking does not aim at achieving an ideal balance that "solves" the paradox but at working through said paradox by embracing tensions.[11]

There is a debate on whether paradoxical tensions are inherent in systems or socially or cognitively constructed, with some scholars referring to the philosophy of social construction for the latter option.[12] Others argue that paradoxes are inherent in the structure of our physical world, for example, in terms of the inherently paradoxical nature of quantum physics.[13] In a somewhat self-referential application of both- and thinking, it would seem likely that the constitution of paradoxical tensions is a combined process based on feedback loops, which involves both social or cognitive construction, and the inherent properties of underlying realities. To illustrate this point, the laws of nature mean that burning fossil fuels unavoidably creates greenhouse gas emissions that fuel global warming and climate change, which is an inherent property of the physical world. At the same time, the decision to burn fossil fuels in the first place is strongly linked to processes of cultural evolution and discovery, as well as the development of social norms that build on socially embedded cognitions of individuals. These processes can potentially stretch over long time periods that involve historical path dependencies but ultimately remain human interpretations, choices, and constructions—even though they can equally create strong inertia.

In paradox theory, a distinction is further made between salient and latent tensions. While tensions between, for example, social, environmental, and economic requirements in the context of sustainable development are (in principle) universal, they only become manifest when "consciously observed, articulated, and experienced by organizational actors."[14] This distinction also matters for businesses' responses to paradoxes and paradoxical tensions. Specifically, proactive responses (having their origins partly in the ambidexterity concept, which refers to an organization's ability to be both efficient in the management of today's business and responsive to possible future demands) address competing demands, such as employees being asked to simultaneously maintain contact with customers and reduce business travel for ecological reasons. Companies can do so by means of temporal or spatial separation of the demands, for example, by means of video conferencing. Proactive responses therefore imply that tensions and trade-offs are made visible and are consciously considered and that contradictions are acknowledged and discussed. Ultimately, proactive responses are mostly seen as the better approach to creating generative solutions to a paradox.

Opposed to this, defensive responses suppress, ignore, or displace one of the two poles (i.e., opposing demands) of the paradox. For example, in the context of sustainability and ecological transformation, companies may only pursue symbolic actions or green-washing rhetoric. As a result, economic demands are implicitly prioritized at the expense of social or environmental concerns, in which case adoption of the latter type of response neglects competing demands, which can then create vicious cycles that result in inertia and paralysis.[15]

Extending the notion of defensive responses, other work has stressed the importance of power dynamics that impact the capacity of individuals to "[…] engage [effectively] with paradoxical tensions."[16] This line of reasoning argues that individuals, such as employees in a company, who are insufficiently empowered are typically missing resources to respond adequately to the contradictory demands that characterize paradoxes and paradoxical tensions. As a consequence, they experience such tensions as so-called pragmatic paradoxes, which lack any acknowledgement of the necessity of competing demands as a precondition for engaging generatively with paradoxes, since some actors have a vested interest in supporting one pole rather than the other. A telling example of such a constellation is the "Dieselgate" scandal, when Volkswagen employees were faced with competing demands (producing sellable cars and meeting exhaust emission legislation) but left without adequate resources and agency to simultaneously address both in a productive manner.[17]

Links of Paradoxes and Tensions with the Role of Innovation for Global Sustainability

In this section, I want to integrate paradox theory, as introduced in the previous section, in more depth with different literatures relating to the environmental humanities in terms of efficiency, consistency, and sufficiency, in order to clarify how tensions and paradoxes in large-scale ecological transformations can be addressed by innovation and technology and where this has limits. This integration attempt as a starting point focuses the technology factor in the IPAT equation, which, as explained before, represents efficiency, and thus ultimately innovation, that enables businesses' efficiency gains. Innovation, in terms of products but also production processes, is a potentially very promising strategy for resolving (even paradoxical) tensions. In the field of economics and management theorizing, innovation leads to so-called creative destruction, that is, the simultaneous creation of new business opportunities and the market exit of firms that become uncompetitive due to not pursuing these opportunities.[18] This process can enable a business case for sustainability beyond pure rhetoric, as exemplified (among others) in discourses about new and more sustainable business models. One example in this regard is manufacturer Patagonia having offered to repair or exchange used jackets, rather than incentivizing consumers to buy more new clothing. Therefore, the mechanisms underlying creative destruction raise a serious question about how paradoxical

"paradoxical" tensions really are, or, stated differently, to what degree technology and innovation via efficiency gains can ultimately resolve grand challenges in the context of sustainable development, as technology optimists would hope.

Innovation economics distinguishes the rate and direction of inventive activity, which can be generalized to innovation and technological development.[19] This means that technological progress toward carbon neutrality may not be necessarily equally conducive for biodiversity, while the invention and commercialization of new computer applications will not always support sustainable development. By this, I mean that climate change in the long run changes habitat conditions globally, which in turn increases the likelihood for species migrating. At the same time, the installation of new renewable energy-generation capacity such as photovoltaics and wind turbines, or the increased adoption of biofuels (all of which mitigate climate change), may also have negative effects on biodiversity. Equally, while digitalization is often associated with positive contributions to sustainability, this depends highly on the specific application case. For example, computer game development is a highly successful IT-based industry, yet scientists raise concerns about negative social effects, especially for adolescents, such as social disconnection or physical health problems.

These examples clarify that, mathematically, innovation can be conceived of as a vector, which helps to explain why a business case for sustainability may or may not emerge from efficiency improvements and other forms of innovation and technological development, depending on what direction the improvement is traveling. In this sense, an innovation (be it from large, established firms, also called "incumbents," or from small start-ups and market entrants) is at most a necessary condition for the aforementioned business case, albeit not a sufficient one.

Economic theory furthermore suggests that more radical innovations are more difficult to realize, not least because incumbents tend to have strong incentives to pursue more incremental innovation.[20] For example, the large-scale transition toward electric mobility was significantly accelerated by automotive firms Tesla and BYD, which newly entered the automotive industry and focused from the outset on electric cars. Therefore, even though incrementalism may be necessary to some degree to enable more radical technological change in the longer term, it holds the risk of short-termism or symbolic action, especially by incumbents, or even worse, it could lead to lobbyism or collusive behavior aimed at stabilizing market positions and related private profits at the expense of social welfare. The Dieselgate scandal is an extreme example of this point, with the incumbent Volkswagen even resorting to illegal means to support its short-term profits, symbolically pretending to meet emission limits in official tests. At the same time, other incumbents in the industry have also used similar means, which is an example of collusive behavior whereby firms, rather than competing for better solutions, coordinate or jointly keep secret the unproductive or even destructive solutions they apply.

Paradox theory enables deeper analyses of these phenomena since the notion of tensions speaks to forms of societal or regulatory framing and framework conditions at different levels that can support or hinder more radical innovation. This has been documented in the case of environmental regulation and for system-level innovation—as needed in the case of large-scale ecological transformations such as transitions to a renewable energy base of industrial and social processes.[21]

In the context of paradox theory, innovation can be understood as a very promising strategy for actively responding to and generatively resolving tensions. Such thinking can be extended to the aspect of consistency, which has been argued to be—at least in some definitional variants—quite alike to efficiency. As explained earlier, consistency is a conceptually very broad term that comprises elements such as high levels of circularity, the usage of substances that already exist in nature, and the avoidance of over-population and resource overuse.

Ultimately, consistency is proposing to make the production and consumption systems of humankind more cyclical and more compatible with natural ecosystems. This is one important basis for requiring that a circular economy is largely bio-based and renewable in terms of materials and energy provision, since it not being sufficiently bio-based would be problematic, for example in relation to issues such as dissipation (i.e., the wide diffusion of human-made substances, such as microplastics or pesticides).

Similarly, while, for example, non-natural composite materials could be used in a circular economy to produce electric cars with carbon fiber bodies, the extensive recycling required, for which almost all energy needs would have to be renewable-based, on a large scale could negatively impact biodiversity and other environmental goods (see my comment above on the relationship between climate change and biodiversity). Therefore, human-made composite materials have only limited compatibility with a circular economy, even though the consistency of human-made substances might increase over time due to technological improvements and innovation. Equally, sufficiency could allow the use of a certain (or higher) amount of human-made materials within a circular economy.

Generalizing this specific case, the views opposed to the win-win logic of a business case for sustainability have linked the disconnect between micro-level progress and continued macro-level damage to the difference in absolute and relative decoupling, as it is termed in the language of ecological modernization.[22] To use again the example of circular economies for illustration purposes, while renewable energies are partly considered a "backstop technology" for supplying a resource at low costs for infinity,[23] this may have some limitations, depending on the scale of implementation.[24] A similar scaling argument applies to the negative investment effects of transitioning to a circular economy from implementing technological innovations that require additional energy and material inputs. Ultimately, therefore, a circular economy will not avoid global limits and issues of absolute decoupling without involving sufficiency.

Put differently, the business case for sustainability (at least in the long term and for the case of profit-oriented firms) implies too high growth requirements to remain within planetary boundaries and limits to growth if it does not integrate sufficiency into business model considerations. For example, corporate attempts to support sustainable development are often confounded by the rebound effects of innovation and novel technologies, as in the case when more energy-efficient appliances increase energy needs due to incorporating more functionality.[25]

Overall, the analysis to this point therefore suggests that next to efficiency, consistency and sufficiency are equally relevant factors for reducing environmental damage, but that the interaction of these three factors can create paradoxical tensions. While this already represents a challenge to solutions, the situation is further complicated by interactions of such tensions across levels, as I detail in the following section. As an initial illustration, a possible example of interaction across levels is seen in the form of globally increasing carbon dioxide emissions and concentrations vis-à-vis sustainability initiatives run by individuals or in firms, cities, regions, industries, or countries.

Paradoxes and Tensions from the Individual to the Global Level

Links between paradox theory and the role of innovation for global sustainability from the perspective of management and economics relate especially to the paradoxical role that companies or organizations are said to play in the context of sustainability.[26] For example, while firms in many ways depend on resources sourced from nature, at the same time—through their operations—they act as a major driver for the decline of biodiversity. These paradoxical tensions for biodiversity reveal interaction across levels, since competing demands exist across different levels for the roles firms are expected to perform.[27] This partly also relates to trade-offs between the short and the long term[28] For example, employees want to save jobs in a company, while the countries in which the company operates want sufficient environmental protection. However, if the latter implies heightened costs, job safety may be under threat, at least in the short term. Yet, in the long term, innovation may reduce costs so that the demands are less contradictory. Similarly, while the global chemical industry has benefited from selling human-made pesticides, society and the environment in many countries have suffered as a consequence.[29] These illustrations demonstrate that paradoxical tensions emerge even more so when a multi-level perspective is taken, for example in terms of countries or regions at the macro-, industries or companies at the meso-, and groups or individuals at the micro-level.[30]

Since it has been proposed that sustainable development is inherently paradoxical, it is important to understand more thoroughly specific instances in this regard, such as biodiversity, particularly, because it has been argued that paradoxes are often latent, only becoming salient through specific processes such as sensemaking,[31] e.g., in our specific illustration due to observable effects and impacts resulting from biodiversity loss.[32] More generally, sensemaking here is understood as an approach in which an actor (which can be, among others, an individual or a firm or an industry) "uses tensions and inconsistencies between considerations to prompt elaboration and clarification of each objective, duty, and concern."[33] Salience is more likely once a multi-level perspective is adopted.

To illustrate this suggestion, the paradox that firms, as an important structural element in many societies, through their core activities harm the context in which they are embedded (in our illustrative case, the natural environment, as exemplified through biodiversity, and thus ultimately also society) is modulated by contextual aspects that lead to more nuanced experiences of how paradoxical tensions become salient. Tensions may vary, for example, across companies or industries at the meso level, and groups or individuals at the micro level, while still interacting between levels. This means that tensions (e.g., those relating to biodiversity or natural capital) may become salient across different levels.[34] Depending on how exactly salience occurs at a specific level, there may be more proactive—or rather reactive—responses to tension. For instance, for firms, tension may become salient in terms of stakeholder pressure, whereas for individuals it may become salient as a feeling of grief when walking in spring through silent landscapes where birds can no longer be heard. Determined further by factors such as the empowerment available to actors at different levels, responses may be more or less generative, even to the extreme that actors experience the tension as a pragmatic paradox.

The heterogeneity across levels discussed in this section is also important for considering tensions across levels, since responding to tensions at only one level may create tensions at another level.[35] Therefore, while the paradox of sustainable development is mirrored by the IPAT equation, the consideration of (paradoxical) tensions across levels reveals heterogeneity—the analysis of which provides some guidelines on how innovation can best address said paradox.

Discussion

Engaging with notions of (potentially paradoxical) tensions concerning innovation and surfacing across levels enables new and more interactive approaches that support necessary comprehensive system transformations, especially in the contexts of climate change, continued global economic and human population growth, and sustainability in a finite world, which due to historic path dependencies and inertia are often very difficult to realize. A paradox lens in this context specifically helps address power issues related to the fact that the realization of sustainable development implies (relative) gains and losses that challenge historically developed and reproduced power distributions and infrastructures. Considering paradoxes and paradoxical tensions in the context of global economic growth (typically resulting from more short-term private benefit or profit opportunities) and long-term sustainability (which ultimately can only be derived from social responsibility concerns) might be further enriched by considering philosophical views on technology.

As a starting point in this respect, Hardensett[36] proposes that modern processes of technology development and innovation require a deep exploration of human nature—much in the way that sufficiency suggests this requirement.[37] While acknowledging an unavoidable human element in technology, Hardensett at the same time strongly advocates for contextualizing technology as being influenced by culture and other social factors relating to the P (population) and A (affluence) factors in the IPAT equation. These considerations almost seamlessly carry over to the work of German philosopher Hans Jonas, who is probably best known for calling for a need to act in a manner that the effects of one's action are compatible with permanent and genuine human life on earth.[38] This so-called ecological imperative builds on a deep philosophical analysis of technological development that stresses positive feedback loops and historic path dependencies that create a technological drift toward forms of innovation that paradoxically simultaneously improve the quality of human life while at the same time deteriorating the state of our planet (in turn negatively affecting conditions for humankind on earth).

Jonas relates this (in line with the interdependency of the factors in the IPAT equation raised above) to the effects of technology increasingly extending in time and space, resulting in those (yet unborn) that have to bear the majority of these effects not being able to influence or demand compensation from those who create and trigger them.[39] The famous World Commission on Environment and Development (WCED) definition of sustainable development as "[...] development that meets the needs of the present without compromising the ability of future generations to meet their own needs"[40] refers to this tension exactly and thus can be considered a (derived and somewhat more operationalized) variant of the ecological imperative itself.

Jonas discusses this point further, arguing that even if nuclear fusion (as another potential backstop technology) were indeed available, the inherent mechanisms and dynamics of technology and innovation would require balancing efficiency gains with consistency and sufficiency to avoid the paradoxical decline of humankind.[41] He derives similar limitations, as discussed above, with regard to renewable energy technologies and, as a consequence, advocates caution and sufficiency, contending further that what is avoidable and which is unavoidable can only be clarified after giving avoidance a serious try.[42] Such thinking would do well as a conceptual elaboration of sufficiency, beyond the operationalization of the latter in the IPAT equation.

Applying the considerations discussed so far, again to the case of the circular economy and its tensions and paradoxes (at different levels), enables us to recognize the need to contribute to efficiency, sufficiency, and consistency jointly. This necessitates, for example, considering detrimental effects to biodiversity in the definition and implementation of renewable energy strategies as one crucial engine for a circular economy (because—as explained above—only a largely bio-based and renewable circular economy appears to achieve necessary levels of consistency in the absence of nuclear fusion being available on a large scale). While initial trade-offs between biodiversity and reducing greenhouse gas emissions may reverse over time and are currently without suitable alternatives in the long term, as laid out above, transition paths still matter because the technological progress best supporting carbon neutrality may differ from the technological progress best supporting biodiversity, and uncertainty makes trading-off between the two types of progress difficult for now.

Even in a sufficiently bio-based and renewable circular economy, a great deal of increased energy and material efficiency is needed to address the backstop technology issue for renewable energy, which may not be resolvable otherwise. Yet, while efficiency thus appears to be one necessary element for the feasibility of a large-scale ecological transformation, consistency and sufficiency seem to additionally be necessary or could even point to an alternative trajectory into a sustainable future. Specifically, advances toward a circular economy exemplify well the definitional elements of a paradoxical tension, namely contradiction, interrelatedness, and persistence. This can be seen from the fact that even in transitioning to a circular economy set up for the absolute decoupling of economic growth from environmental degradation, sufficiency still has a crucial role, at least in terms of buying time in the short term, which corresponds to the caution and sufficient level of frugality advocated by Jonas.[43]

These considerations beg the question as to how to achieve a better balance by building on the various elements of paradox theory. For example, in relation to the salience of tensions, it can be argued that current productivity measures do not account well for increased energy inputs to production systems. Farming productivity, for instance, has seemingly increased, especially as part of the so-called Green Revolution, but in fact the energy intensity of farming has risen a lot, and this can be extended to wider historical evidence on such increased energy inputs, thereby suggesting that more reflective productivity modeling is needed to understand and address the growth implications of apparent efficiency gains, which in reality run counter to consistency and sufficiency efforts.[44]

The potential of sufficiency was involuntarily demonstrated by CO_2 reductions driven by the COVID lockdowns in Germany in 2020. Given this was forced in a much unwanted way, it can be seen as an existence proof of feasible system configurations and at the same time as an indication of the need to proactively coordinate efficiency, consistency, and sufficiency efforts well across levels, to avoid the deprivations and shortages associated with the lockdowns. This is underscored further by some scientists arguing that the COVID pandemic was in effect triggered by human population growth, which, as detailed above, reflects the positive effects of innovation and technological progress.[45]

At the same time, the limitations caused by the COVID lockdowns led to innovation, for example in terms of novel patterns of work, life, and leisure that created new views, perceptions, experiences, and reflections—all of which would seem to hold significant generative potential in the sense of paradox theory, with one example being the large-scale adoption of video conferencing worldwide. Yet, even efficiency and consistency improvements at a scale that enables a "tunnel effect"-style transformation path, e.g., by means of digitalization, would according to paradox theory raise the question whether generative solutions are possible without sufficiency. For example, with regards to digitalization, it has been pointed out that new information technologies

> can […] both be used to speed up the pace of life (work and leisure), thus promoting a society of harried mass consumers, or it can be used to conserve precious natural resources (energy and non-energy) by doing things more intelligently and improving the quality of life without adding stress to the environment[46].

This last distinction is in fact well aligned to Jonas's idea that leisure and abundance possibly paradoxically require each other[47] and also to proposed generative solutions to this tension in terms of prioritizing prosperity, not growth.[48]

Moreover, the abovementioned tunnel effect from efficiency gains remains doubtful from a historic perspective, based on empirical evidence that suggests a tenfold gap in innovation speed, for example with regard to CO_2 efficiency improvements, but also in relation to the efficiency of providing most of the crucial materials in today's increasingly globalized—and thus homogenized—world economy.[49] Moreover, the history of digitalization to date is rather riddled with examples of rebound effects that can be explained by means of their value for addressing the perceived and advertised needs of current generations, albeit at the expense of potentially ignoring the needs of future generations in the sense of the WCED definition of sustainable development. This can be interpreted well through the lens of paradox theory by referencing power effects and pragmatic paradoxes.[50]

As detailed earlier, there is a rate and a direction of innovation activity and, consequently, resulting technological change. As stated in the introduction, this suggests we should ask, under what conditions technological development and creative destruction would lead to firms maximizing efficiency contributions to sustainable development (i.e., what rate and direction the innovative activity of companies should have)? At the core of this question is the need for system innovation, that is, significant, large-scale, synchronized, and system-wide changes that resolve (at least as much as possible) paradoxical tensions between comprehensive changes in technology, the structure of the economy, and sociopolitical and cultural systems.[51] Innovation can therefore only provide sustainable solutions if it is embedded systematically. In other words, while technology initially does not reflect values (in the sense that it can be used in very different directions, not in that it has no historical foundations), once it is involved to achieve a goal (e.g., carbon neutrality), it unavoidably reflects values. In fact, the initial decision to develop technology is often driven by values (which economists would argue are captured by price signals that may be correct or—once externalities exist in underlying markets—biased), which points to vicious cycles and feedback loops that are one core element of paradox theory.

Conclusion

As this chapter aimed to expose, the large-scale transformation of a comprehensive system (integrating technological, economic, socio-political, and cultural sub-systems and encompassing several levels) is riddled with tensions and paradoxes. It starts with a set of values that exist in a society at some point in time (based on historic, idiosyncratic, and path-dependent generation processes). Even if, in this status quo, an agreement can be reached about (more or less) appropriate or necessary choices for the future, implementing them is not always very effective, since it oftentimes meets with paradoxical tension. One reason for this is that sustainable development is in many ways inherently paradoxical and that a reorientation toward a new norm such as sustainable development implies gains and losses (relative to the current situation). This is further complicated by interactions with power distributions and infrastructures that are again historically generated and reproduced in idiosyncratic and path-dependent processes. Westpfahl also alludes to this issue at the end of his contribution when pointing out that large-scale ecological transformations also need to meet social justice requirements.[52] As a result, there is a paradox between a comprehensive transformation of the global economic system and maintaining stability in the day-to-day environment of actors at different levels (such as individuals or companies) or of sub-systems within it.

This also relates back to the question as to whether paradoxical tensions are inherent in systems or socially or cognitively constructed.[53] The generation of paradoxes, as already mentioned, is likely a combined process involving feedback, e.g., in relation to social processes of need construction that imply a tension between the maximum level of freedom for the humans currently living as opposed to ensuring maximum sustainability with regard to later generations, unborn yet. It has been pointed out that this issue is about freedom in the short versus the long term, which can also be referred to as the tension between different forms of what ultimately could be seen as a dictatorship (of the now or the future).[54]

As outlined earlier in the context of the IPAT equation, efficiency is addressed by a technology factor that reflects innovation, sufficiency reflects a willingness to accept reductions in material consumption (when basic needs are well met) and a perception shift at the level of the individual and beyond that such consumption is of lesser relevance relative to immaterial consumption or non-consumption. A paradox lens in this context specifically enables the incorporation of emotional and affective implications and outcomes in order to ultimately support the necessary comprehensive system transformations, especially in the context of climate change and distributional justice, which, due to historic path dependencies and inertia, are often very difficult to realize in present-generation daily life and business. To paraphrase critical theory, all actors can live wrong lives to some degree, since we are faced with historic conditions that are typically not perfectly aligned with current and future demands, but there is a limit to how wrong a life can be lived to remain true (to the facts or other likely longer term trends, such as global change challenging or even threatening humankind). In much the same manner, technology can often be understood in many ways as a history-based negotiation of the future that is affected by a larger or lesser degree by current power distributions.

In the light of what was discussed in this chapter (especially in relation to possible limitations of technology and innovation to increase sufficiency and consistency), paradox theory also provides another avenue to potentially generative responses, namely, that lowering expectations can equally reduce paradoxical (i.e., potentially strong to insurmountable) tensions.[55] Such satisficing clearly is an element of sufficiency and can further help convergence toward a new collective consensus based on changed expectations because it relates voluntary behaviors in support of sustainability at the individual level with higher levels, such as organizations or societies, and in doing so helps to "stay with the trouble"[56] in a manner conducive to the core propositions of both, paradox theory and environmental humanities.

Notes

1 Stephan Schmidheiny, *Kurswechsel* (Geneva: World Business Council for Sustainable Development, 1992).
2 Ann-Kristin Bergquist, "Renewing Business History in the Era of the Anthropocene," *Business History Review* 93, no. 1 (2019): 3–24.
3 . See the chapter by Rafael Hernández Westpfahl, "Mining for a Low-Carbon Economy? Articulations by the Mexican Corporate Sector" in this book.
4 Paul Ehrlich and John Holdren, "A Bulletin Dialogue on 'The Closing Circle': Critique," *Bulletin of the Atomic Scientists* 28, no. 5 (1972): 16–27; Barry Commoner, "A Bulletin Dialogue on 'The Closing Circle': Response," *Bulletin of the Atomic Scientists* 28, no. 5 (1972): 17, 42–56.
5 Jorgen Randers, *2052: A Global Forecast for the Next Forty Years* (White River Junction: Chelsea Green Publishing, 2012).
6 Alexia Panayiotou, Linda Putnam and George Kassinis, "Generating Tensions: A Multilevel, Process Analysis of Organizational Change," *Strategic Organization* 17, no. 1 (2019): 8–37.
7 Wendy Smith, "Dynamic Decision Making: A Model of Senior Leaders Managing Strategic Paradoxes," *Academy of Management Journal* 57, no. 6 (2014): 1592–1623.
8 Marianne Lewis, "Exploring Paradox: Toward a More Comprehensive Guide," *Academy of Management Review* 25, no. 4 (2000): 760–776.
9 Wendy Smith and Marianne Lewis, "Toward a Theory of Paradox: A Dynamic Equilibrium Model of Organizing," *Academy of Management Review* 36, no. 2 (2011): 381–403.
10 Lewis, "Exploring Paradox," 760–776.
11 Lotte Luescher and Marianne Lewis, "Organizational Change and Managerial Sensemaking: Working Through Paradox," *Academy of Management Journal* 51, no. 2 (2008): 221–240.
12 Linda Putnam, Gail Fairhurst and Scott Banghart, "Contradictions, Dialectics, and Paradoxes in Organizations: A Constitutive Approach," *Academy of Management Annals* 10, no. 1 (2016): 65–171.
13 Wendy Smith and Marianne Lewis, *Both/and Thinking* (Boston: Harvard Business Review Press, 2022).
14 Paula Jarzabkowski, Rebecca Bednarek and Jane Lê, "Studying Paradox as Process and Practice: Identifying and Following Moments of Latency and Salience," in *Dualities, Dialectics, and Paradoxes in Organizational Life*, ed. Moshe Farjoun, Wendy Smith, Ann Langley and Haridimos Tsoukas (Oxford: Oxford University Press, 2018), 175–194.
15 Smith and Lewis, "Toward a Theory of Paradox," 381–403.
16 Marco Berti and Ace Volkmann Simpson, "The Dark Side of Organizational Paradoxes: The Dynamics of Disempowerment," *Academy of Management Review* 46, no. 2 (2021): 252–274.
17 Medhanie Gaim, Stewart Clegg and Miguel Pina Cunha, "Managing Impressions Rather than Emissions: Volkswagen and the False Mastery of Paradox," *Organization Studies* 42, no. 2 (2021): 949–970.
18 Joseph Schumpeter, *Capitalism, Socialism and Democracy* (Floyd: Impact Books, 1942).
19 Richard Nelson, "Introduction to The Rate and Direction of Inventive Activity: Economic and Social Factors," in *The Rate and Direction of Inventive Activity: Economic and Social Factors*, ed. Universities-National Bureau Committee for Economic Research & Committee on Economic Growth of the Social Science Research Council (Princeton: Princeton University Press, 1962), 3–16.

20 Kenneth Arrow, "Economic Welfare and the Allocation of Resources for Invention," in *The Rate and Direction of Inventive Activity: Economic and Social Factors*, ed. Universities-National Bureau Committee for Economic Research & Committee on Economic Growth of the Social Science Research Council (Princeton: Princeton University Press, 1962), 609–626.

21 Michael Eugene Porter, "America's Green Strategy," *Scientific American* 264 (1991): 96; Frank Boons and Marcus Wagner, "Assessing the Relationship between Economic and Ecological Performance: Distinguishing System Levels and the Role of Innovation," *Ecological Economics* 68, no. 7 (2009): 1908–1914.

22 Armin Grunwald, "Diverging Pathways to Overcoming the Environmental Crisis: A Critique of Eco-Modernism from a Technology Assessment Perspective," *Journal of Cleaner Production* 197, no. 2 (2018): 1854–1862.

23 Amnon Levy, "From Hotelling to Backstop Technology," Department of Economics, University of Wollongong, 2000, accessed March 27, 2023, https://ro.uow.edu.au/commwkpapers/25.

24 Ted Trainer, "Can Renewables etc. Solve the Greenhouse Problem? The Negative Case," *Energy Policy* 38 (2010): 4107–4114.

25 Mona Chitnis et al., "Turning Lights into Flights: Estimating Direct and Indirect Rebound Effects for UK Households," *Energy Policy* 55 (April 2013): 234–250; Mona Chitnis et al., "Who Rebounds Most? Estimating Direct and Indirect Rebound Effects for Different UK Socioeconomic Groups," *Ecological Economics* 106 (2014): 12–32.

26 Connie Van der Byl and Natalie Slawinski, "Embracing Tensions in Corporate Sustainability: A Review of Research from Win-Wins and Trade-Offs to Paradoxes and Beyond," *Organization & Environment* 28, no. 1 (2015): 54–79.

27 John Lorimer Campbell, "Institutional Analysis and the Paradox of Corporate Social Responsibility," *American Behavioral Scientist* 49, no. 7 (2006): 925–938; Jason Jay, "Navigating Paradox as Mechanisms of Change and Innovation in Hybrid Organizations," *Academy of Management Journal* 56, no. 1 (2013): 137–159; Joshua David Margolis and James Patrick Walsh, "Misery Loves Companies: Rethinking Social Initiatives by Business," *Administrative Science Quarterly* 48, no. 2 (2003): 268–305.

28 Natalie Slawinski and Pratima Bansal, "Short on Time: Intertemporal Tensions in Business Sustainability," *Organization Science* 26, no. 2 (2015): 531–549.

29 Simone Müller, "Hidden Externalities: The Globalization of Hazardous Waste," *Business History Review* 93, no. 1 (2019): 51–74.

30 Panayiotou, Putnam and Kassinis, "Generating Tensions," 8–37.

31 Karl Edward Weick, "The Collapse of Sensemaking in Organizations: The Mann Gulch Disaster," *Administrative Science Quarterly* 38, no. 4 (1993): 628–652.

32 Olivier Boiral and Inaki Heras-Saizarbitoria, "Managing Biodiversity through Stakeholder Involvement: Why, Who, and for What Initiatives?" *Journal of Business Ethics* 140 (2017): 403–421.

33 Margolis and Walsh, "Misery Loves Companies," 291.

34 Putnam, Fairhurst and Banghart, "Contradictions," 65–171; Margolis and Walsh, "Misery Loves Companies," 268–305.

35 Smith and Lewis, "Toward a Theory of Paradox," 381–403.

36 Heinrich Hardensett, *Philosophie der Technik* (Marburg: Metropolis, 2017).

37 While Kapp is often cited as one of the most eminent scholars in this context, his view of organ projection (i.e. the notion that all technology is an extension of human organs) as the root of modern technology and innovation is too limiting and narrow. Hardensett

pointed to this well before World War II but for political reasons could not publish his ideas until much later.

38 Hans Jonas, *Das Prinzip Verantwortung: Versuch einer Ethik für die technologische Zivilisation* (Frankfurt/M.: Insel-Verlag, 1979).

39 Ibid., 56.

40 WCED (World Commission on Environment and Development), *Our Common Future* (New York: Oxford University Press, 1987), 54.

41 Jonas, *Das Prinzip Verantwortung*, 324–328.

42 Ibid., 378.

43 Ibid., 319–329.

44 Robert Ayres and Benjamin Warr, *The Economic Growth Engine: How Energy and Work Drive Material Prosperity* (Cheltenham: Edward Elgar, 2009).

45 Mark Everard et al., "The Role of Ecosystems in Mitigation and Management of Covid-19 and Other Zoonoses," *Environmental Science and Policy* 111 (2020): 7–17.

46 Daniel Spreng, "Possibility for Substitution Between Energy, Time and Information," *Energy Policy* 21, no. 1 (1993): 13–23.

47 Jonas, *Das Prinzip Verantwortung*, 319.

48 Tim Jackson, *Prosperity without Growth: Foundations for the Economy of Tomorrow* (London: Routledge, 2017).

49 Jackson, *Prosperity without Growth*, 96–101; Timothy George Gutowski et al., "The Energy Required to Produce Materials: Constraints on Energy-Intensity Improvements, Parameters of Demand," *Philosophical Transactions* 371, no. 1986 (2013): 1–14.

50 Berti and Simpson, "The Dark Side of Organizational Paradoxes," 252–274.

51 Chris Freeman, "History, Co-Evolution and Economic Growth," *Industrial and Corporate Change* 28, no. 1 (2019): 1–44; Boons and Wagner, "Assessing the Relationship," 1908–1924.

52 Raphael Hernández Westpfahl, "Mining for a Low-Carbon Economy?"

53 Putnam, Fairhurst and Banghart, "Contradictions," 65–171.

54 Hans Joachim Schellnhuber, *Selbstverbrennung – Die fatale Dreiecksbeziehung zwischen Klima, Mensch und Kohlenstoff* (München: C. Bertelsmann, 2015).

55 Xin Li, "Dialogue: Solving Paradox by Reducing Expectation," *Academy of Management Review* 46, no. 2 (2021): 406–419.

56 Donna Haraway, *Staying with the Trouble: Making Kin in the Chthulucene* (Durham: Duke University Press, 2016).

5 Toxic Commons and the Politics of Ambivalence

Re-imagining Toxic Legacy Sites

Simone M. Müller and Angeliki Balayannis

Introduction[1]

The grazing bison at the Rocky Mountain Arsenal National Wildlife Refuge, with the skyline of Denver as their backdrop, were an iconic sight, marrying multiple strands of human–environment relationships throughout US history. Covering more than half a page in the TRAVEL section of the *Roanoke Times* and other smaller newspapers that had bought the Associated Press's package, the image was placed to evoke readers' desire to experience nature the American way.[2] Only a ten-minute drive from downtown Denver, visitors to the refuge could "step into nature and see the native wildlife."[3] This included a potential meet and greet with the last descendants of a species that a century-and-a-half earlier had been hunted by white American colonizers almost to extinction. As part of a national restoration and rewilding program, a small bison herd was moved into Rocky Mountain Arsenal in 2007 (see Figure 5.1).[4] On its website, the park advertises different routes for different time schedules, among them a one-hour, 11-mile self-guided drive compatible with a specially recorded podcast, i.e., low-motion access to "native wildlife" while remaining in the heart of Denver.[5] Like the bison, national parks are iconic institutions in the United States, signifying the country's long-standing commitment to the protection of nature on top of a heated and fundamentally colonial debate on what counts as "wilderness."[6] National parks have been forged through a complicated, one could say *ambivalent*, relationship between nature and culture, or rather between notions of roadlessness and the sacrality of this wilderness, on the one hand, and an expanding automobility of and accessibility to nature, on the other.[7]

Yet, even in this context of cultivated wilderness, Rocky Mountain Arsenal is unique among US wildlife refuges. The 15,000-acre site is a former military installation, making the ambivalence even more pronounced. In the 1940s and 1950s, it hosted a chemical weapons factory for the US Army and then was leased to companies producing commercial pesticides. Before it became a wildlife refuge, the area was part of the US superfund program dedicated to cleaning up and remediating the country's most contaminated sites.[8] As a prior location for military activities, a veil of secrecy still lies over much that had been going on at Rocky Mountain Arsenal. Nonetheless, it is undisputed that some of the wildlife

DOI: 10.4324/9781032627984-7

Figure 5.1 Rocky Mountain Arsenal, four bison with a Denver skyline.

Source: © Elisa Dahlberg/US Fish and Wildlife Service, October 13, 2016.

refuge is still heavily contaminated.[9] To this day, the U.S. Fish and Wildlife Service fences off no-go areas in the park and prohibits swimming, wading, and bathing in all refuge waters, alongside the use of boats, canoes, kayaks, float tubes, paddle boards, and other flotation devices.[10] Some of the "pristine" nature is too toxic to behold.

Toxic legacy sites such as Rocky Mountain Arsenal are a systemic—yet often sidelined—issue in the processes that this book, "Ecological Ambivalence, Complexity, and Change: Perspectives from the Environmental Humanities," calls the "ecological transformation of society." Toxic legacy sites, both figuratively and physically, illustrate how societies across the world are confronted with millions of metric tons of toxic chemicals, both as synthetic and natural materials, that have over centuries been released into our air, water, and soil as part of industrial, agricultural, and military activities. They represent how the past lingers in the present and can harm (generations of) human and more-than-human bodies in substantive and yet unequal ways. For a long time, they have primarily figured as an engineering issue aimed at the restitution of "pristine" nature with an away-place, an ultimate sink, where extracted contaminated soils could be taken to vanish.[11] Nonetheless, as toxic materials can never disappear, they can only be "transformed, transmuted, combusted, combined or any other manner of material change."[12] How, then, do we deal with what remains?

It is the responsibility of the Environmental Humanities to pose questions (and to find answers) relating to how to live with a toxic past that remains in the present.

Recently, researchers, activists, and artists have started to shift their attention from future-oriented clean-up and remediation processes to discourses and practices of how people form (and have formed) relationships with already existing toxic legacies in the present. Under the framing of toxic heritage, a small but growing number of scholars and artists are working on the intersection of how (and if) sites of environmental harm are treated as "heritage," and how this heritage accounts for environmental harm.[13] Artistic-research interventions, meanwhile, seek to theorize a deepened understanding of a planet experiencing multiple crises while reaching out to a larger, non-academic public about toxicity, coloniality, and unequal burdens.[14]

Joining forces of a historian and a human geographer and approaches from environmental history with feminist STS, our contribution forges new conceptual pathways in this burgeoning field. Pivoting from critiques of chemical exposures and toxic entanglements toward interventions and experiments that can confront these legacies,[15] this contribution introduces the concept of *Toxic Commons* to respond to questions relating to how to work with complex and unruly legacies such as Rocky Mountain Arsenal. Toxic Commons is both a heuristic category for historicizing the processes of the increasing contamination of our air, water, and soil and an analytical category, a relational epistemology, that challenges notions of clean-up and containment and seeks to invoke commons literature to inspire a new ethics of care for the toxic. Ultimately, Toxic Commons—a form of "negative" commons—invites a politics of ambivalence that seeks to respond to a past that lingers in the present and to a state of contamination that divides as much as binds us.[16]

Studying Toxic Commons beyond Remediation and Clean-up

Toxic legacy sites are notoriously difficult to study. Scholars in the social sciences and humanities have long demonstrated the legal, material, and political barriers to this research[17], consequently making it challenging to recognize links across sites and their shared processes and institutions. For example, different national legislations make it difficult for researchers to engage holistically and across nation states with toxic heritage sites—an issue that often commences with geolocating. While in the United States, in Austria, or in Switzerland, the toxic legacy cadasters help identify specific sites, the German *Atlastenkataster* obscures exact location markers. Prioritizing data security, particularly if legacy sites are still classified as suspicious sites, German agencies only give away the village or town and the kind of usage leading to contamination.[18] Often overgrown with grasses, old waste deposits, for instance, are difficult to see without technoscientific interventions. Old industrial sites, in turn, tend to be fenced off, giving the air of inaccessibility and thus at least the illusion of toxic containment—an illusion that is often nourished from proud narratives of industrial heritage.[19]

Another issue involves assigning responsibility for sites of contamination and damage and thereby establishing, if not a juridical, then a historical lineage to help understand the site as a legacy formed through materiality as well as relations. Polluters are often states that are no longer in existence or companies that have dissolved and abandoned or lost their archives. It is difficult to individualize responsibility to consumers as pollution emerges at sites of manufacturing or disposal, and causality is often impossible to determine because it is too expensive within existing governance mechanisms, such as a lack of "return to sender" programs.[20] The expansive temporalities and spatialities of these sites ultimately demonstrate the limits of the law, notions of responsibility, and regulatory mechanisms. As the chapter will detail further below, toxic legacy sites exceed political and even human timescales, with geographies that are unruly, thus transgressing boundaries and borders.[21] The way they have been archived and cared for in the past, finally, makes it difficult to develop an organigram of the relationships toxic legacy sites entertain with various actors and institutions.

In response, scholarship has often turned elsewhere, with "elsewhere" in this regard meaning the attempt to rectify the past by looking into the future. Even before the UN acclaimed its "Triple Crisis" of 2020, pairing contamination with climate change and biodiversity loss, research on planetary health, remediation, and clean-up vastly expanded with studies seeking how to restore a pristine planet.[22] Dominant forms of mainstream (and often white) environmentalism, meanwhile, have focused on calls to stop pollution at its source. The new global treaty to address plastic waste "from source to sink" is the clearest example of this mode of environmental politics.[23] Calls to halt the production of new, *future* pollution—and to shift to "circular" or "zero waste" economies—are of course crucial for addressing extraction.[24] As we like to broaden the attention from the chemicals and policies of clean-up to practices and discourses resonating with today's ubiquity of toxic contamination and the commonality of exposure, we prefer to speak of and work with a present that is shaped and informed by a past that lingers on in polluted soil, rivers, buildings, bodies, and memories. We do so by introducing the Toxic Commons.[25]

Toxic Commons is both a heuristic category, as in the age of the Toxic Commons, for historicizing the process of the increasing contamination of our common-pool resources—our air, water, and soil—and with it an increasing, albeit unequal, commonality of toxic exposure since at least industrialization. It is also an analytical category, a relational epistemology, that seeks to reconcile the situated with the planetary in the context of pollution as a global problem, and with this some of the ambivalences of ecological transformation. As this chapter will illustrate, Toxic Commons challenges notions of containment and invokes commons research to also include negative externalities. Furthermore, through them, we seek to enable ways of understanding, living with, and ultimately caring for toxic legacies, thereby responding to a lack of adequate epistemological, ethical, and technological tools to intervene at toxic sites. Moreover, it invites researchers to move beyond a recognition of being-in-common to investigate discourses and participate in practices of commoning, i.e., sharing knowledge and decision-making about toxic sites and their legacies in ways that go beyond the individual or the singular community.

Introducing the Age of the Toxic Commons

Toxic Commons, that is, heuristically speaking, contaminated soil, air, or water, have built up over the course of centuries. As historians can trace from archival sources, and geologists, earth scientists, and toxicologists from soil or rock samples, some of the earliest contaminated sediments go back to mineral mining across Asia, the Americas, and Europe. Gold, iron, copper, and tin were mined, smelted, and worked into ornaments, utensils, and weapons in Southeast Asia long before the Common Era, nourishing active trade between China and the larger Southeast Asian region that expanded rapidly. By the 17th century, the Chinese also exploited mineral resources in what is today Burmese and Lao territory before the British took over the exploitation of these resources and their trade. In the German Ruhr valley, in the Upper Harz, and similarly in Lower Hungary, miners have extracted coal for over five centuries, not only sinking the land, in some cases by up to 60 feet, but also blowing millions of tons of coal dust into the air. In the Americas, indigenous groups mined for gold in the Caribbean, Mexico, and the Andes, while Spanish mining efforts in one of the largest silver deposits at the Cerro Rico de Potosí, today Bolivia, commenced as early as 1545, followed by other mining endeavors in what are today Peru, Brazil, Colombia, and Chile. In all cases, mining activities have resulted in a historical legacy of heavy metal contamination, particularly antimony and arsenic in the surrounding areas.[26]

With industrial capitalism evolving first in Europe and then the rest of the world from the 18th century onward, pollution patterns, their scale, and how they have been breached and understood fundamentally changed. On the back of a shift in energy regimes from solar to fossil fuels, new technologies, ranging from the steam engine to generators and the steam turbine, unleashed industrial power in unprecedented ways. New and faster mobility and communication patterns in turn nurtured the built-up of a global market structure catering to appetites for ever more diverse products, all of which resulted in almost endless pollution and waste streams, with rivers, lakes, and creeks running black and air suffocatingly orange, as vividly captured by Claude Monet and other painters of the London Smog or the Black Country, an industrial region in the center of England.[27] The 20th century saw the emergence of toxic legacy sites on an even grander scale, due to industrialization and agriculture's growing intensity, their globalization, and an array of new contaminants. Production processes that relied increasingly on synthetic chemicals while being, at least until the 1970s, vastly unregulated in terms of discards and disposal, a vastly expanding and globalizing consumer culture, and a high-production-yield agriculture based on an excessive use of pesticides, all fed into vast waste and pollution streams. Scientists and environmentalists, ranging from Rachel Carson to Barry Commoner, alerted us to topics including pesticides and waste incineration.[28] Additionally, the use of chemical warfare in the First World War, as well as on a grand scale in Vietnam and Iraq, and the continued extraction of energy resources and primarily fossil fuels have greatly added to the creation of a permanently polluted planet.[29]

The tonnage, ubiquity, and longevity of toxic chemicals within, but also beyond, toxic legacy sites have led some scholars to describe this planetary condition as a "new age of toxicity" or a "ubiquitous condition of chemically altered living-being."[30] While this classification has placed an increasing spotlight on contamination as a key issue of concern, it also symbolizes the growing obfuscation of our understanding of "the toxic." In the last two decades at least, the boundaries of what constitutes a pollutant have become more porous with, so far, poorly understood cumulative effects due to the longevity of chemicals' molecular makeup and the emergence of a cocktail of different toxicants. While yesterday's persistent organic pollutants, such as DDT, are still present, hundreds of thousands of new chemicals continue to be marketed.[31] Today, there are over 350,000 synthetic human-made chemicals in the world that have accumulated over decades, if not centuries.[32]

Additionally, as toxicants perceptible to the senses, such as fumes and blackened water, no longer constitute the greatest danger from toxic legacy sites, attention has shifted to bodies that are saturated with toxic substances.[33] There is also growing evidence that many synthetic chemicals can cause serious health problems, including cancer, neurological problems, and endocrine disruption. Since the 1990s, and as highlighted by Theo Colborn's 1996 publication *Our Stolen Future*, endocrine-disrupting chemicals, i.e., compound chemical agents that can alter a body's hormonal system, have come to the fore, invisibilizing and complicating modes of exposure.[34] With endocrine-disrupting chemicals, public health specialists have had to extend their metric for ascertaining harm switching from dose to moments of exposure and the possibility of multiple exposures, also called cocktail effect.[35] Toxic Commons have arrived on the small scale of the human and more-than-human body, as well as the grand scale of the planet. A study warns that the Earth's threshold for chemical pollution has been breached, meaning that its impacts on systems such as nitrogen cycling or the spread of invasive species are beyond our capacity to understand, let alone control.[36]

In this history of accumulating contamination with overlapping modes of exposure, we are cognizant that the creation of a permanently polluted planet is a product of racial capitalism. The term "environmental racism" stands at the beginning of the environmental justice movement, recognizing how minoritized groups are disproportionately affected by toxic pollution.[37] Unequal exchanges and geographically expanding externalization processes are also an integral part of global pollution patterns, thereby constituting one of the downsides of modern environmentalism, in that its nationalistic approach has created an environmental governance framework through which usually white elites of the Global North live well at the cost of others far distant, where less strict rules and regulations apply.[38] Since the 1980s, the global trade in hazardous waste material has benefitted from such differences, making the export of toxicants from North to South an economic opportunity.[39] Since 2007, the NGO Pure Earth has assembled lists of the world's most contaminated sites through its Toxic Site Identification Program (TSIP), with Sub-Saharan Africa and South and Southeast Asia consistently ranking first in this regard. Alongside the relocation of hazardous waste from North to South, risk and the outsourcing of dirty production are also—intentionally—unevenly distributed.[40]

Toxic Commons as Unruly and Uncontainable

Thinking with Toxic Commons as an analytical category involves a fundamental shift away from binary conceptualizations of contamination and cleaning up, questioning notions of territoriality and containment. The concept asks scholars to challenge the intricate linkages between environmental protection and national sovereignty as enshrined by the United Nations system after the Second World War and reemphasized by the institutionalization of modern environmental governance since the 1970s.[41] Instead, Toxic Commons suggests embracing the ambivalence that comes with recognizing contamination, particularly on a planetary scale, as unruly and often uncontainable. The point we are making also begs for the recognition of a historical trajectory based on the understanding of a planetary materialism that has thus far received little attention in the established narrative of modern environmentalism, namely that the long-range atmospheric transport of pollutants, as in acid rain or radionuclear clouds, or similarly long-range transports in water, as in the Great Pacific Garbage Patch, have played a fundamental role in the creation of the Toxic Commons.

Scientists and politicians commenced in the early 1980s to work around phenomena that they termed "problems without passports," almost all of which were linked to issues of contamination.[42] A first step in that direction was the completion of a shift toward a public understanding of planetary circularity, not only in global trade but also in atmospheric terms. Scientists were puzzled as they continued to measure the same pesticides banned in the U.S. a decade earlier in the Great Lakes and other bodies of water.[43] In Europe, meanwhile, foresters, politicians, and society discussed the issue of *Waldsterben*, eventually connecting it to acid rain, i.e., the precipitation of sulfuric or nitric acid, with suspected sources located hundreds of kilometers away.[44] Coming through the air, nitric or sulfuric acids originating in the United Kingdom, for instance, traveled in global circulations of particulate clouds before rain washed them into different landscapes in Central and Northern Europe. While the topic of long-range atmospheric transport had entered a scientific, niche discourse with a particular focus on mercury, the *Waldsterben* discourse boomeranged the issue as a top international priority. By 1983, the topic had gained such political momentum that it led to the passage of the UN Convention on Long-Range Transboundary Air Pollution.[45]

Not only atmospheric circularity but also trade patterns made the commonality of contamination a felt reality. For instance, for the first time, in 1981, environmental activists cited the failures of national governance for the containment of toxic chemicals in the context of a free-market-driven global economy.[46] Pesticides, such as DDT, chlordane, or heptachlor, which the U.S. Environmental Protection Agency had taken off the market because they were seen as carcinogenic and capable of endangering humans, wildlife, and the environment, continued to be exported to other parts of the world where they were still legal.[47] As an unintended consequence, they also returned to the U.S. through the global foodways of tropical fruits, coffee, tea, meat, or corn. In 1986, the U.S. General Accounting Office reported that the U.S. Food and Drug Administration (FDA) had found substantial pesticide residues in a number of commodities in the three years prior.[48] In an 1988 case, for example, chlordane levels in beef imported from Honduras were eight times those allowed in the United States.[49] By the end of the 1980s, this "Circle of Poison" also led American legislators to pass a hapless amendment of the Federal Insecticide, Fungicide, and Rodenticide Act (FIFRA).[50]

The reactor catastrophe at Chernobyl, the industrial accident at Bhopal, and the extremities of the Ozone Hole, all neatly happening in close proximity to each other, next emphasized how fervently, violently, and completely harmful substances disregarded logics of containment and easily transgressed political, social, and technological boundaries set up for their control. In 1986, people around the world fearfully watched the catastrophe of nuclear meltdown stretching itself easily far beyond what was then the Soviet Union.[51] In 1984, the Indian city of Bhopal provided a blueprint for an industrial accident when a silent and invisible toxicant indiscriminately killed children and grandparents, women and men, blue and white collar workers, simply because they were within a certain radius of the explosion.[52] Witnessing the rapid thinning of the planet's protective shield, the ozone layer, finally and vividly illustrated the global dimension of borderless chemical threats.[53]

In response to the recognition of these processes, public and scientific international discourses grew from the 1980s onward around the transboundary nature of environmental catastrophe that also informs, albeit partially, our conceptualization of Toxic Commons. Published shortly after the Chernobyl accident, Ulrich Beck's *Risk Society*, later followed by *Ecological Politics in an Age of Risks*, appeared to capture the mood of the time perfectly.[54] Beck saw modernity enter a new phase under the sign of technologically produced risks, in which it was now less about the distribution of the wealth produced and more about the distribution of the risks. This, in turn, led to a completely new social constellation: "Necessity is hierarchical, smog is democratic," Beck argued.[55] His thesis of a democratization of pollution has been heavily contested based on the vast differences in people's opportunities to mitigate effects from exposure, the impoverishment of entire city districts, or the existence of suburban retreat areas.[56] Yet, his point on the leveling effects of an increasingly ubiquitously polluted planet still rings true, as does the underlying realization that in the early 21st century, there are next to no pristine, clean environments that can be neatly zoned off from a contaminated majority world.

Additionally, other conceptual approaches to apprehending the complexity and reach of pollution as a global problem have proliferated in the last few decades, first highlighted by the UN's 1987 landmark publication *Our Common Future*.[57] Introducing the concepts of sustainable development and intergenerational justice, the report extended frameworks of governance, remediation, and responsibility from the past and the present into the future.[58] Around the same time, activist movements forged the term and notion of global environmental justice as a key political response to draw links across seemingly disconnected situations and to highlight the transboundary violence of capitalism and colonialism.[59] The global waste trade and the unequal distribution of hazards connected to it became one of global activists' key issues in the late 1980s, with the United Nations' Basel

Convention on the Transboundary Movement of Hazardous Wastes and Their Disposal as one of the first international governance frameworks to work around environmental justice.[60]

In the last decade, the Anthropocene has been one of the most influential reconfigurations, with geoscientists re-thinking humanity's ecological impact and its more-than-human entanglements in geologic time—an epochal shift. Here, the human is imagined as a species with the potential to fundamentally affect planetary processes, including biospheric, atmospheric, and hydrospheric disruptions.[61] More recently, the health impacts of the Anthropocene have been increasingly approached through an interdisciplinary "planetary health" lens that grew out of "global health" initiatives emerging in the 1990s, cutting across the disconnected logics of public, environmental, and global health.[62]

The social sciences and humanities have responded to the Anthropocene with forceful critiques of thinking at the scale of the planetary, adding a number of different -ocenes as alternatives.[63] They critique the continuation of centering the human; erasing the unevenness and "patchiness" of global environmental problems; and insufficiently targeting capitalism and colonialism as drivers of environmental damage.[64] In Chernobyl and Bhopal, the effects of pollution from these industrial accidents have unfolded along the lines of class, caste, race, and gender—and continue to do so in ways that are institutional and embodied. In the context of the ozone hole, particular regions of the world and particular societal groups from within these world regions were less able to escape potential harm.[65] To return to Michelle Murphy, toxicity is an ambivalent condition; it divides as much as binds us.[66]

Invoking the Negative Commons

Thinking with Toxic Commons as an analytical category involves revisiting the commons and unlocking its potential for an ethics of care. Rather than Ulrich Beck or the Anthropocene, we find in economist Elinor Ostrom a productive source with whom to think when it comes to toxic legacy sites, thereby inspiring the concept of Toxic Commons. Commons-based research is flourishing in the environmental social sciences.[67] Thinking with commons—that is, jointly owned resources—and through commoning practices—that is, self-organized and needs-oriented joint production, administration, or use of these resources—seems to open up a "new paradigm that embodies its own logic patterns of behavior, functioning as a different kind of operating system for society."[68] Commons scholarship sees the commons as potential new models for societies' organization and (ecological) transformation in opposition to a capitalist world organized around profit-oriented entrepreneurs and the nation state, concepts of growth and private property, thus taking inspiration from scholars like Naomi Klein or activists like David Bollier who see the commons, in contrast, as a "new paradigm," a "coalition of coalitions," or "a silent revolution."[69]

Practices of commoning are key to this new paradigm, i.e., acts of (multi-species) relations "made under the expectation that we will take care of one another" (and the planet) and with "a shared understanding that some things belong to all of us."[70] Nevertheless, the literature remains fixed on the commons as a positive. In our work, we extend Ostrom's approach of communal governance and responsibility to negative externalities as commons. As neoclassic models of economy have seen nature and its resources as being free of cost, the pollution of common-pool resources, such as air, water, and soil, has become a negative externality outside of economic modeling, which, in turn, presupposes private property.[71] In Toxic Commons, externalities are reintroduced, but not as private property or as an entity that itself holds private property but as a universal good that belongs to all of us.[72] The question we seek to evoke through Toxic Commons is what happens if we apply Hugo Grotius' idea of Mare Liberum to toxic legacy sites, if these, too, become a shared heritage of humankind instead of a *res nullius*?[73] Toxic legacy sites, we argue, belong to all of us, too.

We argue that this expansion of commons to Toxic Commons challenges what commoning means, as well as who is involved and on what scale. The investigation of toxic commoning as shared practices of governance and responsibility for toxic legacies is about neither flattening power relations nor reducing actions to the community level, leaving affected communities to sort things out on their own. Toxic legacies, i.e., contaminated and damaged sites and relations, transgress sectoral and geographical boundaries, necessitating the intervention of actors beyond the affected community, which subsequently calls for a multi-scalar and often, given the nature of multinational entrepreneurship, international stewardship for these sites and affected communities.

The history of the internalization of pollution is one floored with failures. And yet, our endeavor is not to be read as a renewal of Hardin's "Tragedy of the Commons."[74] Ostrom's work has effectively answered popular theories around Hardin's essay, which postulated that private property is the only way to protect finite resources from depletion. Additionally, we argue, notions of individual ownership of toxic legacies, reducing responsibility to the community level, follow the same flawed, colonial, and racist logics.[75] Rather, as Ostrom illustrates, successful collective forms of governance rely on clear rules, frameworks, and common institutions that involve everyone concerned and affected.[76]

Conclusion

What is described herein is in many ways a politics of ambivalence. Bringing commoning and toxicity together enables a form of pragmatism and experimental ethos that moves beyond "purity" politics that demand clean slates.[77] This is not only a call to "make do" on a damaged planet—following a long line of scholars in the environmental humanities[78]—but also a recognition that disposal, remediation, or "clean-up" are always, partially, *ambivalent*. The damage cannot be undone. Which brings us to the question: if environments and bodies remain forever irrevocably contaminated, what constitutes a "good" response? Toxic Commons takes this question as its starting point. It is an analytical tool to foreground material processes of local persistence and planetary diffusion that make exposure to contamination an increasingly common and communal experience. Furthermore, it facilitates research that engages with the failures of national ecological containment and the limits to safeguarding national environments and citizens in a highly mobile world of global markets. Ultimately, though, it serves as a counter-narrative to established histories that legitimize the state as a primary actor and a respondent of intervention and ecological transformation.[79]

Rocky Mountain Arsenal is a vivid example not only for the ambivalences of ecological transformation but also for the challenges involved in transforming toxic legacy sites into a community asset and effort, what we would call a Toxic Commons. It is a transformation, however, that requires more than just the clean-up: "More than in a physical sense," such as a kind of ecological transformation, "requires a shift in mindset and acceptance by the community" concluded law professor Rachael E. Salcido after closely following transformation processes at Rocky Mountain Arsenal.[80] In 1992, US Congress passed the Rocky Mountain Arsenal National Wildlife Refuge Act to make space for transforming the military site into a wildlife refuge. The Act restricted the sale of part of the land still containing waste and remediation facilities and defined that the vast acreage was to be managed by the Federal Fish and Wildlife Service. In 2004 and 2006, 4,930 and 7,266 acres of land, respectively, were transferred from the Army to the Department of Interior, to be managed by the Federal Fish and Wildlife Service. The bison returned to Rocky Mountain Arsenal in 2007. Meanwhile, the State of Colorado and the US Army were locked in a court battle over whether the former could assert authority. The situation worsened when, in 2001, an unexploded M-74 munition was found and the Army and its contractors decided to put it back where it was found and to deceive the Colorado Department of Public Health and Environment regarding their discovery.

At first site, Rocky Mountain Arsenal National Wildlife Refuge may appear as a Toxic Commons failure. During the years of the initial clean-up, as Salcido relates, the Denver community had little opportunity to contribute to or engage in restorative acts because of the dangers of pollution at the site and a lack of trust in ensuring the safety of entrants. Moreover, at times, the community has been "enraged by the lack of transparency and outright deception perpetrated by those responsible for restoration, and frustrated by marketing of the refuge which downplayed the risks and trade-offs made in the remedy selected for the site."[81] Overall, the ecological transformation of the Rocky Mountain Arsenal has been marked by a "community that has been alternatively deceived, shut out of decision-making and had its concerns marginalized," and yet it remains a community that has experienced "skepticism, fear, and *hope for the future* all at once."[82] Today, the Denver community is living with Rocky Mountain Arsenal as its neighbor. Is this enough of a success? We think so, at least for now.

Notes

1 On citation and interdisciplinarity: an endnote on endnotes. When writing this contribution, citation style became a way of thinking about what it means to interdisciplinary

research. Disciplinary differences between history and geography—for us, this is effectively between the social sciences and humanities—materialized in citational practice. Angeliki has no experience of working with (or learning how to use) endnotes, whereas for Simone, they are central to historical research practice. This difference is more than a matter of style. It raises questions around what matters in research, or even what *constitutes* it. For Angeliki, in-text citation quickly points to specific concepts and approaches; a conceptual shorthand. It also directly recognizes the work of others in her text—and this co-presence of author and citation is, for Angeliki, about citational politics. For Simone, the endnote, and even more so the footnote is a pivotal space for recognizing the specificity of sources, including often very lengthy primary source citations and their interpretation. In history, interpretation outranks frameworks, as the latter, particularly if they are rigorously defined, often do not stand the test of "change over time." In-text citations, for Simone, can be disruptive for the reader and get in the way of storytelling, which has become a key mode of research within the environmental humanities, and is how we arrived at our decision to endnote. Learning from Max Liboiron's reimagination of the capacities and ethics of footnotes and endnotes, we also approach this space as an opportunity to discuss our interdisciplinary practice.

2 Donny Bryson (Associated Press), "Mile-high Haven," *Roanoke Times*, November 4, 2012.

3 U.S. Fish and Wildlife Service, "*Rocky Mountain Arsenal National Wildlife Refuge*," last accessed December 12, 2023, www.fws.gov/refuge/rocky-mountain-arsenal; Andrew C. Isenberg, *The Destruction of the Bison. An Environmental History, 1750–1920* (Cambridge: Cambridge University Press, 2000).

4 Rachael E. Salcido, "The Rocky Mountain Arsenal National Wildlife Refuge. On a Rocky Road to Creating a Community Asset," *John Marshall Law Review* 47, no. 4 (2018): 1399–1430, here 1407.

5 U.S. Fish and Wildlife Service, "*Rocky Mountain Arsenal National Wildlife Refuge*."

6 William Cronon, "The Trouble with Wilderness, or getting back to the wrong nature," in *Uncommon Ground: Rethinking the Human Place in Nature*, ed. William Cronon (New York: Norton, 1996), 69–90.

7 Paul S. Sutter, *Driven Wild, How the Fight Against Automobiles Launched the Modern Wilderness Movement* (Seattle: University of Washington Press, 2005).

8 On the US Superfund Program, see Joel A. Mintz, *Enforcement at the EPA: High Stakes and Hard Choices,* Revised Edition (Austin: University of Texas Press, 2012).

9 Salcido, *The Rocky Mountain Arsenal*, 1400.

10 U.S. Fish and Wildlife Service, "*Rocky Mountain Arsenal National Wildlife Refuge*".

11 Joel Tarr, *The Search for the Ultimate Sink: Urban Pollution in Historical Perspective* (Ohio: University of Akron Press, 1996).

12 Angeliki Balayannis, "Toxic Sights: The Spectacle of Hazardous Waste Removal," *Environment and Planning D EPD: Society and Space* 38, no. 4 (2020): 772–790, here 772.

13 For an example from the humanities, see: Elizabeth Kryder-Reid and Sarah May (eds.), *Toxic Heritage Legacies, Futures, and Environmental Injustice* (London: Routledge, 2024); Caroline Ektander and Jonas Stuck, "Redistribute Toxicity," *Environmental Humanities* 15, no. 3 (2023): 104–118; Gustav Wollentz et al., "Toxic Heritage. Uncertain and Unsafe," in *Heritage Futures. Comparative Approaches to Natural and Cultural Heritage Practises*, eds. Harrison Rodney et al. (London: UCL Press, 2020), 294–312.

14 In 2018, Antonia Alampi, Ayushi Dhawan, Caroline Ektander, Maximilian Feichtner, Simone M. Müller, and Jonas Stuck founded the artist-research network *Toxic Commons*

realizing two joint exhibitions, e.g., Caroline Ektander and Antonia Alampi, *The Long Term You Cannot Afford: On the Distribution of the Toxic* (Berlin: SAVVY Contemporary, 2019), exhibition catalog, accessed February 3, 2023, https://savvy-contemporary.com/site/assets/files/5558/toxicity_handout.pdf.

15 Angeliki Balayannis and Emma Garnett, "Chemical Kinship. Interdisciplinary Experiments with Pollution," Catalyst. Feminism, theory, technoscience 6 No. 1 (2020): Special Section on Chemical Entanglements: Gender and Exposure, 1–10, https://doi.org/10.28968/cftt.v6i1.33524.

16 Michelle Murphy evokes this quote of being divided and yet bound together in her article, yet not quite in the same context. "Alterlife and Decolonial Chemical Relations," *Cultural Anthropology* 32, no. 4 (2017): 494–503.

17 R. Shiloh Krupar, *Hot Spotters Report: Military Fables of Toxic Waste* (Minneapolis: University of Minnesota Press, 2013).

18 For the United States, see US EPA, "National Priority List and Superfund Alternative Approach Sites," last modified June 17, 2024, www.epa.gov/superfund/search-superfund-sites-where-you-live. For Switzerland, see Swiss Environmental Protection Agency, last modified December 4, 2019, www.bafu.admin.ch/bafu/de/home/zustand/daten/geodatenmodelle/altlasten--geodatenmodelle.html; for Austria, see Austrian Environmental Protection Agency, "Altlastenatlas," www.altlasten.gv.at/atlas/verdachtsflaechenkataster.html. For Germany, see the Altlastenkataster, organized by each federal state; for instance, Bavaria: www.lfu.bayern.de/altlasten/altlastenkataster/altlastenauskuenfte/index.htm.

19 Marina Weinberg and Valentina Figueroa, "Politics of Mining. Toxic Heritage in the Atacama Desert," in *Toxic Heritage Legacies, Futures, and Environmental Injustice*, eds. Elizabeth Kryder-Reid and Sarah May (London: Routledge, 2024), 109–123.

20 Angeliki Balayannis, "Toxic Sights," 772–790.

21 Simone Müller and May-Brith Ohman Nielsen (eds.), *Toxic Timescapes. Examining Toxicity across Time and Space* (Ohio: Ohio University Press, 2023).

22 For example, National Research Council, Division on Engineering and Physical Sciences, Board on Army Science and Technology, Committee on Review of the Conduct of Operations for Remediation of Recovered Chemical Warfare Materiel from Burial Sites, *Remediation of Buried Chemical Warfare Materiel* (Washington, D.C., 2012); Leo van Velzen (ed.), *Environmental Remediation and Restoration of Contaminated Nuclear and NORM Sites* (Cambridge: Woodhead Publishing, 2015).

23 United Nations Development Program, "A global treaty to end plastic pollution is in sight," November 2023, last accessed December 14, 2023, www.undp.org/blog/global-treaty-end-plastic-pollution-sight.

24 The literature on both circular economy and zero waste concepts is too vast to cover, for a start: Sabine Oberhuber and Thomas Rau, *Material Matters. Wie eine neu gedachte Circular Economy uns zukunftsfähig macht* (Berlin: Econ, 2021); Pablo del Río et al., *The Circular Economy. Economic, Managerial and Policy Implications* (Cham: Springer, 2021); Robert Crocker and Steffen Lehmann, *Designing for Zero Waste: Consumption, Technologies and the Built Environment* (London: Earth Scan, 2012).

25 Simone M. Müller, "Toxic Commons: Toxic Global Inequality in the Age of the Anthropocene," *Environmental History* 26 (2021): 444–450; Angeliki Balayannis, "Toxic Sights," 772–790.

26 On mining history and contamination, see Johann Paul, *Die Erft im Spannungfeld von Bergbau und Landwirtschaft. Einwirkungen auf ein rheinisches Flussgebiet vom 17. bis ins 20. Jahrhundert* (Siegburg: Rheinlandia Verlag, 1999); Franz-Josef Brüggemeier,

Michael Farrenkopf, and Heinrich Theodor Grütter (eds.), *Das Zeitalter der Kohle. Eine Europäische Geschichte* (Essen: Klartext, 2018); John R. McNeill and George Vrtis, *Mining North America. An Environmental History since 1522* (Oakland: University of California Press, 2017); Nicholas A. Robins, *Mercury, Mining, and Empire. The Human and Ecological Cost of Colonial Silver Mining in the Andes* (Bloomington: Indiana University Press, 2011); Anthony Reid et al., "Chinese on the Mining Frontier in Southeast Asia," in *Chinese Circulations. Capital, Commodities, and Networks in Southeast Asia*, ed. Eric Taglioacozzo and Wen-chin Chang (Durham: Duke University Press, 2011); Tina Asmussen, "Eisen war ich… Kupfer bin ich, Gold werd ich," *Hypotheses (open edition) Materialized Histories. Materielle Kultur und digitale Forschung* (2021), https://mhistories.hypotheses.org/4470; Peter Bakewell, "Mining in Colonial Spanish America," in *The Cambridge History of Latin America. Vol. 2, Colonial Latin America*, ed. Leslie Bethell (New York: Cambridge University Press, 1984), 105–152.

27 Anita Lam, "Choking to Death. True Crime and the Great Smog," in *The Routledge Handbook of Crime Fiction and Ecology*, ed. Nathan Asham (New York: Routledge, 2023).

28 Rachel Carson, *Silent Spring* (Boston: Mifflin, 1962).

29 The literature on these chemical processes is too vast to consider herein. For an overview, consult Müller and Ohman Nielsen (eds.), *Toxic Timescapes*; François Jarrige and Thomas Le Roux, *The Contamination of the Earth* (Cambridge: The MIT Press, 2020) or Benjamin Ross and Steven Amter, *The Polluters. The Making of Our Chemically Altered Environment* (Oxford: Oxford University Press, 2010). On pesticides, see Michelle Mart, *Pesticides, a Love Story. America's Enduring Embrace of Dangerous Chemicals* (Kansas: University of Kansas Press, 2015); Frederick Rowe Davis, *Banned. A History of Pesticides and the Science of Toxicology* (New Haven: Yale University Press, 2014). On waste, see Simone Müller, *The Toxic Ship. The Voyage of the Khian Sea and the Global Waste Trade* (Seattle: Washington University Press, 2023); Roman Köster, *Müll. Eine schmutzige Geschichte der Menschheit* (Munich: C.H. Beck Verlag, 2023). On the military aspect, see Jacob Darwin Hamblin, *Arming Mother Nature. The Birth of Catastrophic Environmentalism* (Oxford: Oxford University Press, 2013); Krupar, *Hot Spotter's Report*. David Biggs, *Footprints of War. Militarized Landscapes in Vietnam* (Seattle: University of Washington Press, 2018).

30 Brett L. Walker, *Toxic Archipelago: A History of Industrial Disease in Japan* (Seattle: University of Washington Press, 2010), xi; Murphy, "Alterlife," 494–503, here 497.

31 Jarrige and Le Roux, *The Contamination of the Earth*, 300.

32 Zhanyun Wang et al., "Toward a Global Understanding of Chemical Pollution: A First Comprehensive Analysis of National and Regional Chemical Inventories," *Environmental Science Technology* 54, no. 5 (2020): 2575–2584.

33 Stacy Alaimo, *Bodily Natures. Science, Environment, and the Material Self* (Bloomington: Indiana University Press, 2010); Nancy Langston, *Toxic Bodies* (New Haven: Yale University Press, 2010).

34 Theo Colborn, *Our Stolen Future* (New York: E. P. Dutton, 1996).

35 May-Brith Ohman Nielsen, "Living with Poison. Exploring Generations as Toxic Timescapes," in *Toxic Timescapes. Examining Toxicity across Time and Space*, eds. Simone Müller and May-Brith Ohman Nielsen (Ohio: Ohio University Press, 2023), 21–48.

36 Linn Persson et al., "Outside the Safe Operating Space of the Planetary Boundary for Novel Entities," *Environmental Science Technology* 56, no. 3 (2022): 1510–1521.

37 Robert D. Bullard, "Confronting Environmental Racism in the 21st Century," *Race, Poverty & the Environment* 10, no. 1 (2003): 49–52.

38 Simone Müller, "Hidden Externalities. The Globalization of Hazardous Waste," *Business History Review* 93, no. 1 (2019): 51–74; Stephan Lessenich, *Living Well at Others' Expense: The Hidden Costs of Western Prosperity* (Cambridge, UK: Polity Press, 2019); Marco Armiero, *Wasteocene, Stories from the Global Dump* (Cambridge: Cambridge University Press, 2021).

39 Müller, *The Toxic Ship.*

40 Bret Ericson et al., "Approaches to Systematic Assessment of Environmental Exposures posed at Hazardous Waste Sites in the Developing World: The Toxic Sites Identification Program," *Environmental Monitoring and Assessment* 185, no. 2 (2013): 1755–66.

41 Richard N. L. Andrews, *Managing the Environment, Managing Ourselves: A History of American Environmental Policy* (New Haven: Yale University Press, 2006); David Larsson Heidenblad, *The Environmental Turn in postwar Sweden: A New History of Knowledge* (Lund: Lund University Press, 2021); Perrin Selcer, *The Postwar Origins of the Global Environment: How the United Nations built Spaceship Earth* (New York: Columbia University Press, 2018).

42 On the governance aspect, see Thomas Weiss, *Governing the World? Addressing Problems without Passports* (Boulder: Paradigm Publishing, 2014).

43 Nancy Langston, *Sustaining Lake Superior: An Extraordinary Lake in a Changing World* (New Haven: Yale University Press, 2019).

44 U.S. Congress, Office of Technology Assessment, "Acid Rain and Transported Air Pollutants: Implications for Public Policy" OTA-O-204 (Washington, 1984); Arne Kaijser, "Combatting 'Acid Rain': Protecting the Common European Sky," in *Greening Europe: Environmental Protection in the Long Twentieth Century – a Handbook*, ed. Anna-Katharina Wöbse and Patrick Kupper (Berlin, Boston: De Gruyter Oldenbourg, 2022).

45 T. A. Jackson, "Long-Range Atmospheric Transport of Mercury to Ecosystems, and the Importance of Anthropogenic Emissions: A Critical Review and Evaluation of the Published Evidence," *Environmental Review* 5, no. 2 (1997).

46 David Weir and Mark Schapiro, *Circle of Poison: Pesticides and People in a Hungry World* (San Francisco: Institute for Food and Development Policy, 1981). Greenpeace, "Exporting banned pesticides: Fueling the circle of poison. A case study of Velsicol Chemical Corporation's Export of Chlordane and Heptachlor" (1989).

47 Alice Crow, "Breaking the Circle of Poison: EPA's Enforcement of Current FIFRA Export Requirements," *The Georgetown International Environmental Law Review* 4, no. 319 (1992).

48 U.S. General Accounting Office, "Pesticides: Better Sampling and Enforcement Needed on Imported Food" GAO/RCED-86-219 (1986).

49 Greenpeace, "Exporting banned Pesticides," 21.

50 Crow, "Breaking the Circle of Poison".

51 Kate Brown, *Manual for Survival: A Chernobyl Guide to the Future* (New York: W.W. Norton & Company, 2019).

52 Suroopa Mukherjee, *Surviving Bhopal* (New York: Palgrave Macmillan, 2016).

53 Stephen O. Andersen and K. Madhava Sarma, *Protecting the Ozone Layer* (London: Earthscan, 2002).

54 Ulrich Beck, *Risk Society. Towards a New Modernity* (London: Sage Publications, 1992; Ulrich Beck, *Ecological Politics in an Age of Risk (Cambridge: Polity Press, 1995).*

55 Beck, *Risk Society*, 48.

56 Frank Uekötter, "Die Katastrophe im Zeitalter ihrer elektronischen Reproduzierbarkeit, oder: Wege zu einer Historisierung der Risikotechnologien," *Werkstatt Geschichte* 63 (2013): 101–107.

57 World Commission on Environment and Development, *Our Common Future* (Oxford: Oxford University Press, 1987).

58 Iris Borowy, *Defining Sustainable Development for Our Common Future* (Abingdon: Routledge, 2013).

59 David Naguib Pellow, *Resisting Global Toxics* (Cambridge: MIT Press, 2007).

60 Müller, *The Toxic Ship.*

61 Literature on the Anthropocene, both in the natural sciences and the environmental humanities, is vast and rapidly growing. A good introduction from the Environmental Humanities is Helmut Trischler, "The Anthropocene", *NTM* 24 (2016): 309–335, or Clive Hamilton, Christophe Bonneuil, and François Gemenne (eds.), *The Anthropocene and the Global Environmental Crisis* (London: Routledge, 2015). For environmental history responses, see Christophe Bonneuil, Jean-Baptiste Fressoz, and David Fernbach, *The Shock of the Anthropocene: The Earth, History and Us* (London and New York: Verso Books, 2016) or Dipesh Chakrabarty, *The Climate of History in a Planetary Age* (Chicago and London: University Press of Chicago, 2021).

62 Eduardo Missoni, Guglielmo Pacileo, and Fabrizio Tediosi, *Global Health Governance and Policy. An Introduction* (London: Taylor and Francis Group, 2019).

63 There is, for example, the Capitalocene, the Wasteocene, the Plantationocene, the Chlutlocene. See Jason W. Moore (ed.), *Anthropocene or Capitalocene? Nature, History, and the Crisis of Capitalism* (Oakland: PM Press, 2016); Armiero, *Wasteocene*; Donna Haraway, "Anthropocene, Capitalocene, Plantationocene, Chthulucene: Making Kin," *Environmental Humanities* 6, no. 1 (2015): 159–65.

64 Chakrabarty, The Climate of History, 2021; Timothy Morton, *Hyperobjects. Philosophy and Ecology after the End of the World* (University of Minnesota Press, 2013). Kregg Hetherington, "Introduction: Keywords of the Anthropocene", in *Infrastructure, Environment, and Life in the Anthropocene*, ed. idem. (Durham: Duke University Press, 2019).

65 Müller, "Toxic Commons".

66 Murphy, "Alterlife".

67 Elinor Ostrom, *Governing the Commons. The Evolution of Institutions for Collective Action* (Cambridge UK, Cambridge University Press, 1990); Elinor Ostrom, *Understanding Institutional Diversity* (Princeton University Press, 2005).

68 David Bollier and Silke Helfrich (eds.), *The Wealth of the Commons* (Amherst: Levellers Press, 2012), xi.

69 Naomi Klein, "Reclaiming the Commons," *New Left Review* 9, no. 9 (2001): 81.

70 Peter Linebaugh, *The Magna Carta Manifesto* (Berkeley: University of California Press, 2008).

71 Fredrik Albritton Jonsson and Carl Wennerlind, *Scarcity. A History from the Origins of Capitalism to the Climate Crisis* (Cambridge: Harvard University Press, 2023); Roger Perman (ed.), *Natural Resource and Environmental Economics* (Harlow and Munich: Pearson Education Limited, 2011).

72 On commons and property rights, see Carsten Herrmann-Pillath, "The universal commons: An economic theory of ecosystem ownership," *Environmental Economics* 208 (2023): 1–11.

73 Hugo Grotius, *Mare Liberum* (1609); With Toxic Commons we are explicitly not invoking the notion of a res nullius that belongs to no one, e.g., Karin Mickelson, "The

Maps of International Law: Perceptions of Nature in the Classification of Territory," *Leiden journal of international law* 27, no. 3 (2014): 621–639.

74 Garrett Hardin, "The Tragedy of the Commons," *Science* 162, no. 3859 (1968): 1243–1248.

75 Ralph Ghoche and Unyimeabasi Udoh, "A Commons for Whom? Racism and the Environmental Movement," in *Transforming Education for Sustainability*, ed. Maria S. Rivera Maulucci, Stephanie Pfirman, and Hilary S. Callahan (eds) (Cham: Springer, 2023), 75–88.

76 Andrew P. Follett, Brigham Daniels, and Taylor Peterson, "The Tragedy of Garrett Hardin's Commons," in *The Cambridge Handbook of commons research innovations*, ed. Sheila R. Foster and Chrystie Flournoy Swiney (Cambridge: Cambridge University Press, 2021).

77 Max Liboiron, *Pollution is Colonialism* (Durham: Duke University Press, 2021); Alexis Shotwell, *Against Purity: Living Ethically in Compromised Times* (Minnesota: University of Minnesota Press, 2016).

78 Anna Lowenhaupt Tsing et al., *Arts of Living on a Damaged Planet: Ghosts and Monsters of the Anthropocene* (Minneapolis: Minnesota University Press, 2017).

79 See, for instance, Oliver Schutter and Tom Dedeurwaerdere, *Social innovation in the service of social and ecological transformation. The rise of the enabling state* (London: Routledge, 2022).

80 Salcido, "The Rocky Mountain Arsenal," 1399–1430, here 1402.

81 Ibid., 1412.

82 Ibid., 1412.

Part II

Ambivalences in Practice

6 In Praise of Ambivalence

Reflections on Experiences of Pollution and Remediation

Sebastian Lundsteen

Introduction

It is December 6, 2020, and Bjarne Hansen, a 70-year-old retired fisherman, is sitting in his living room in Thyborøn, a small town in Western Jutland, Denmark, surrounded by cameras, journalists, and technicians. He is in the midst of giving an interview to a regional Danish broadcasting company, explaining how *it* is utterly ludicrous and that *it* has exceeded the thresholds by several thousand, when his phone suddenly rings. He pauses, looks down and up again, and then asks the interviewer if it would be all right if he takes the call. Setting the phone on speaker while the filming continues, he answers, "Yes, it's Bjarne." "Hi, Bjarne," a female voice emerges. "Hi, Pia…!" Bjarne cuts in, while the voice on the other end struggles to complete the sentence, "… it's Pia Olsen-Dyhr." Bjarne's interruption shatters the illusion of unmediated television reality, as if the TV crew accidentally happen to be in his home when the leader of the third biggest political party in Denmark, the Socialist People's Party, calls him.

Obviously knowing what is coming, Bjarne, in a giddy manner, momentarily seems unable to control his movements, and thus he unwillingly juggles his phone. Pia Olsen-Dyhr almost shouts, "It is finally happening!"—a statement provoking a very emotional response from the retired fisherman, who laughs and nearly bursts into tears while erratically rubbing his earlobe. "We will find all the money to clean up Høfden," she continues. Bjarne's eyes become remarkably moist while he laughs excitedly. Then he asks, "How much money did we get?" Silence follows another moment of silence. Pia Olsen-Dyhr titters as if she does not understand his question. After a short break, Pia informs Bjarne that they have received all the money required to clean up the chemical depot "*Høfde 42-depotet*" and begin remediation on a different but related site. After the call ends, the journalist witnessing the interaction asks Hansen how he feels. His facial expression immediately changes to a more serious state while his gaze reveals an intense moment of reflection. "I need time," he responds.[1]

When Bjarne Hansen received the staged call from the politician Pia Olsen-Dyhr in 2019, it happened against the backdrop of what had been touted as a historic climate election. The election demonstrated how green politics had made it to the top of the agenda, superseding popular topics like the EU and immigration,

DOI: 10.4324/9781032627984-9

thus making environmental themes particularly politically decisive.[2] The Social Democratic Party seized power by promising large-scale environmental and climate action and by forming an alliance with Olsen-Dyhr's Socialist People's Party, which vehemently supported one-party government. One of Olsen-Dyhr's irrevocable demands was to map, list, and remediate Denmark's ten most contaminated sites, known as the "Generational Pollutions." The initiative was equivalent to the American Superfund Sites program, running since the 1980s, as the national government's attempt to clean up its industrial legacies.

The three most polluted sites in Denmark are found in Harboøre Tange, a small peninsula in Western Jutland. Of the three contaminated locations, "*Høfde-42 depotet*" is arguably the most urgent and (in)famous depot—a submerged depository on the coastline facing the North Sea[3] (see Figure 6.1). It is estimated to contain more than 120 tons of highly toxic chemical waste from different organophosphate-based products, mainly mercury and residues from the pesticide Parathion.[4] Bjarne Hansen, who has lived on the peninsula all his life, is part of an environmentalist group, "*Høfde-42 Gruppen*," an intergenerational constellation of zealots who have raised awareness and lobbied for remediation for decades. The group's resistance to the pollution and the polluter, namely, the Danish chemical corporation Cheminova, can be traced back to the 1950s.

While the decision to finance the remediation of the site had succeeded much to their credit, a few weeks after the call from Olsen-Dyhr, the group announced that its work was far from finished; in fact, it had only just begun.[5] As such, the

Figure 6.1 The long arm extending into the sea is a "groin,"—a coastal construction that mitigates erosion. The chemical depot is buried next to the 42nd groin.

Source: Copyright Niels Åge Skovbo, Fokus Foto.

environmentalists, spearheaded by Hansen, did not drop their activities after announcing the remediation. Rather, they shifted focus from lobbying to monitoring the clean-up process, which they did via social media communication, collecting data, and holding meetings with politicians, engineers, and other experts. In this way, the group did not reconcile with the promise of a clean-up but signaled a vivid form of skepticism—outright ambivalence—toward the remediation project.

This story is about Bjarne Hansen, *Høfde-42 Gruppen*, and the hazardous waste depot.[6] Hansen is the main figure in the group and embodies it in many ways. He has been called a "pain in the neck" by the local community, politicians, and the media for his stubborn and relentless insistence on removing the chemical waste that still saturates the area. Hansen has also gained national fame, subsequently earning him heaps of accolades, such as a knighthood, for fighting his cause.[7] But Bjarne is not the first in his family to receive such an honor. In 1995, his father, Aage Hansen, received the title for the same thing—a lifelong battle against the pollution at Harboøre Tange.[8] When in 2019, after years of lobbying, funding for the remediation was secured, Bjarne Hansen knew they had won. However, he did not feel like winning. Something did not feel right. In fact, he had been ambivalent about the entire process leading up to remediation.

I wish to think about ambivalence in the context of environmental urgency by sitting with the lack of enthusiasm shown by Hansen and *Høfde 42 Gruppen*. I aim to explore ambivalence as an attitude that arises in processes of ecological transformation, such as remediation, which is informed by the longevity of activists' fight to succeed with their demands, historical instances of failure, and uncertainties concerning the proper documentation of toxic materials.

The article makes three interrelated arguments. First, ambivalence does not exclusively signal rejection or ineptitude but is a particular—and a very strong—attachment to a given situation. People who are ambivalent about something are often so for specific, frequently historically informed reasons. Second, ambivalence is a way of expressing complexity, thus expanding and contesting knowledge regimes, as it articulates submerged knowledges of past experiences that carve out other spaces for expertise rendered illegible in techno-scientific argumentation and economic reasoning. However, it does not oppose these scientific domains but views them skeptically in the claim for monopoly. Third, I hold that ambivalence makes an important contribution to the field of environmental justice, in that it elucidates specific attitudes supporting enviro-political conflicts and signifies how to re/act in a world where, according to historian Michelle Murphy, "industrially produced chemicals have become a part of living-being."[9] In this way, ambivalence potentially addresses otherwise hidden power structures and histories of injustice expressed through internalized frustration. Moreover, it emerges when communities are promised prosperity but instead receive intergenerational contamination and ongoing environmental violence.

Furthermore, the inherent convoluted nature of ambivalence is a crucial yet understudied aspect of ecological transformation that calls for an investigation tracing its roots to understand how and why the attitude emerges. Additionally, it contributes to contemporary conversations relating to environmental humanities by acknowledging multiplicity, complicity, and complexity while insisting that it is not a default or the only response.

Approaching Ambivalence from the Environmental Justice Context

From climate anxiety to anger-induced activism, an increasing body of research is occupied with affective responses to current global dilemmas.[10] Moreover, the quest to find an adequate answer to scenarios often inscribed with dystopian tropes remains part of an ongoing discussion wherein each position appears troublesome. The ambivalent position, however, has eluded attention as a way of acting in a "permanently polluted world" that is also facing the rapid onset of a climate catastrophe,[11] perhaps because ambivalence appears irrational, or perhaps because it designates conflict rather than harmony.

Ambivalence, as defined by psychologists and in psychoanalysis, is considered a single emotion among several but still includes two or more attitudes to a given thing or process, foregrounding contradiction or "mixed feelings."[12] Philosopher Hili Razinsky argues that contrary to conventional depictions of ambivalence as irrational, it is a tension grounded in the desire to resolve a conflict, which is, in fact, rational.[13] Anthropologists Ciara Kierans and Kirsten Bell challenge Razinsky's position by arguing that ambivalence should not be considered a form of tension, in need of resolution, or irrational. By attending to the relationship between the individual and the structural, they contend that "[...] ambivalence is [...] something produced by (and productive of) our orientation to the social world, rather than constituting an explanatory device in and of itself."[14] Thus, anthropologists reclaim ambivalence as a heuristic device, meaning that "we have to accept that things are not readily clear, that we need to learn about the conditions [...] and about how people come to take up variable positions."[15] As such, ambivalence is a path via which we can explore the structures producing the attitude.

Resonating with the structural, the field of Environmental Justice investigates the often negative and discriminatory systemic and political dynamics shaping socio-ecological transformation and community resistance to power.[16] Shifting the attention from isolated and intentional acts of environmental injustice to the structural provides an analytical opening illuminating the production of—and the experience of living with—environmental damage across time and space.[17] However, relatively little attention is paid to the complex and paradoxical attitudes individuals experience when subjected to structural environmental discrimination.

Notably, anthropologist Deana Jovanović has looked at the ambivalence of living in a sacrifice zone that supplies one's livelihood, while Marien González-Hidalgo explores environmental movements' struggles with being a part of the same hegemonic power structures they oppose.[18] Regarding historical investigations, Julie Sze foregrounds the importance of studying past environmental movements to illuminate how historical and cultural forces and resistances emerge by examining the paradoxical relationship between a dire present and hope for a better future.[19] These contributions provide clear-cut distinctions between right and wrong, nuanced perceptions of agency, and showcase how environmental struggles are—and have always been—complex.

In this chapter, I offer an additional perspective by showing how ambivalence operates as a discursive configuration resembling a counterbalance to hegemonic narratives expressing and articulating submerged knowledge as a mode of narrative justice. If discourses on green transitions are saturated with rationality, thereby separating emotions from techno-scientific reasoning, as eco-feminist philosopher Val Plumwood argues,[20] ambivalence is an interjection. Furthermore, it is not solely a response to a given situation or an individual attitude; rather, it is a way of engaging with and problematizing historically contingent injustices that continue to unfold to this day. While the illogical and contradictory nature of ambivalence is excluded or framed as a problem and opposes areas dictated by regimes of action and urgency, it still confronts an otherwise coherent narrative occupied by (techno-scientific) rationality.

Contesting techno-optimistic discursive monopolies is critical in the struggle for a more just world, and storytelling is one affordance in this regard. According to Donna Houston, it is a crucial tool for environmental justice movements, in that it is "a framework for understanding how multiple realities of environmental injury come together in ways that are not always readily discernible through policy or scientific practice."[21] In this sense, environmental justice also consists of narrative strategies but is expressed in myriad ways. I take ambivalence as one strategy employed to unearth historical trajectories and counter a one-sided demand for accepting progress in whatever shape it might come. Following Donna Haraway, who insists that "it matters what stories we tell, to tell other stories with," ambivalence not only tells a story, but it also offers a particular way of telling it.[22] As such, ambivalence is the performative modality of a story, not just the story itself, and it expresses a way of "staying with the trouble" by avoiding deterministic and teleological narratives while acknowledging multiple differentiated and situated perspectives.[23] As articulated by environmentalists, ambivalence expands discursive fields, each desiring remediation while simultaneously being suspicious of the many power structures and interests shaping such processes.

The Making of an Away

To the authorities and the local population in Harboøre Tange, their relationship with "*Høfde-42 depotet*" was ambivalent from the very outset. Following its establishment in the 1950s, the hazardous waste depot was riddled with contestation about placement, what it contained or might contain, and whether the waste was safely confined. Making an "away" for chemical waste generated tension between repeated depictions of a safe and controlled space while ongoing breaks and leaches suggested otherwise, which transformed into ambivalence decades later.

Cheminova arrived in Harboøre Tange in Western Jutland in 1952 from Copenhagen after encountering increasing pressure from neighbors and authorities as a result of chemical pollution.[24] Amidst a series of court trials, the company moved far away from the attention of the Danish capital. The new location was situated on a narrow artificial peninsula that separated two bodies of water, namely, the North Sea to the west and the enclosed water of Limfjorden to the east. Since discharge into water bodies was relatively common in the 1950s, the new location, surrounded as it was by water, provided Cheminova with waste disposal possibilities,[25] and so the company embarked on charming the local population and politicians with promises of prosperity, which were particularly enticing, as the area struggled with employment because of a declining fishing industry and had yet to experience infrastructural development.[26]

Producing organophosphate-based pesticides—a relatively novel agricultural chemical family—meant that Cheminova was also a scientific endeavor, producing its own expertise on the substance and subsequent pollution. Additionally, in 1944, the company was donated to Aarhus University, which immediately established a research fund as formal owners. By creating this type of ownership structure, Cheminova's financial surplus was channeled into societal and immaterial spheres. As such, a transubstantiation of capital was exchanged between a private corporation and a public knowledge-producing institute.[27] This peculiar private–public arrangement formed a connection between expertise and ongoing contestation about chemical pollution.

For example, when Cheminova relocated to Harboøre Tange with a documented history of pollution, the company had to gain permission to dispose of waste as a condition for its production. At the time, a regional monitoring health commission facilitated regulation in collaboration with the police as the enforcer. However, the general directive was formulated vaguely and did not address chemical pollution, while competence was lacking locally.[28] A local court assembly handled permissions to discharge sewage, but as requested by Cheminova, the Ministry of Interior, the Ministry of Fishing, and the Health Ministry, a commission consisting of judges and environmental experts, advised the court to meet local concerns about Cheminova's contamination.[29] At the hearings, a board member and prominent Professor of Law at the University of Aarhus led Cheminova's negotiations. The final agreement restricted discharges to liquid excess, while the most acutely toxic liquid waste had to be disposed of in a drainage area between two dunes on the shoreline facing the North Sea, next to the 42nd groin.[30] Despite initial and grave concerns about the advent of a known polluter, the compromise demonstrated how Cheminova, with the help of legal expertise, successfully managed to shift focus—from authorization to pollute to the area best for polluting.

To alleviate local concerns about contamination, Cheminova promised to establish a pipeline directly into the North Sea, which would, like the ever-rising smoke from 19th-century factory chimneys, give the appearance of sending pollution far away. However, the company reported to the authorities that this plan was technically and financially unfeasible.[31] Instead, the Coastal Department, which held jurisdiction over the area, and Cheminova came to a mutual agreement, namely, a truck transporting sewage into the drainage area. Both agreed that the North Sea made the toxic wastewater harmless through dilution, while percolation removed the rest.[32]

The arrangement proved difficult for minimally treated sewage (if treated at all), while toxic liquid waste filled most of the drainage area and seeped into other spots. Spillovers from open carriers and scalar mismatches between the amount of waste and the area's capacity collectively created a toxic cesspool. Moreover, the makeshift waste infrastructure caused aggravation for the local community to the extent that they had no option but to file charges.[33] The Cheminova's success created an accumulating problem of amassing hazardous waste that did not match the slower temporality of permeation through the granular and dense sand. Neither did it confine to the limited space of the designated area and, instead of creating a controlled drainage area, a toxic lake emerged.

The drainage area was perfect for its assimilative capacity, a term that, according to Max Liboiron, legitimizes pollution through epistemic regimes that measure how much an environment can hold before damage occurs. As such, assimilative capacity denotes how much harm an environment can take before it becomes perceptible.[34] In addition, assimilative capacities hinge on the belief that land and the environment are stable; however, in Harboøre Tange, harm emerged in many forms and temporalities: either disruptive during storms, the provision of incorrect information, or spills from overfilled trucks, or delayed and prolonged, appearing as medical symptoms decades later.[35] Furthermore, the chemical waste seldom stayed in place but spilled and overflowed, thereby contradicting estimations based on static perceptions of the environment and thus creating ambivalence as it became unmanageable.

In 1956, Cheminova urged the Coastal Department to dispose of solid waste next to this toxic soup. Despite being an illegal and unsustainable solution, the company had deposited solid waste on its premises, and now the property was full.[36] Cheminova CEO Gunnar Andreasen argued that there was no other option than to deposit the solid waste in the dunes, adding that, in time, the toxic sewage would dissolve the solids.[37] Andreasen found the drainage basin an ingenious system because of its simplicity: storms would "lick" clean the space between the two dunes, taking it to the North Sea and eventually diluting it in the North Atlantic.[38] Cheminova was reluctant to give any information about the character or the amount of waste, and the Coastal Department discovered that the company had already disposed of contaminated materials next to *Høfde-42*. Recognizing it was probably too late, the seeping area became a toxic landfill from 1957 to 1962.[39]

Cheminova was allowed to dispose of its waste through different strategies: threatening to move its business to less environmentally regulated countries, providing unclear information about the amount and level of toxicity, guaranteeing minimal environmental damage, or promising to improve waste management.[40] Furthermore, the waste posed a crucial dilemma for the Ministry of Interior, as it already existed and was most likely not a problem that would disappear: was it best to have Cheminova managing its own toxic waste by storing it on its own premises but being less monitored, or would an authorized landfill mean increased control? Based on this reasoning, the dunes were considered a better sink for controlling waste.[41]

In 1962, the sink and the volatility of the coastal environment became too perilous an arrangement when a winter storm resulted in breaking the front dike that was supposed to shelter the landfill from the sea. The storm flushed out the liquid waste, which circulated throughout the local area, amounting to an environmental catastrophe and forcing the authorities to prohibit dumping. Subsequently, the landfill was covered with a sand lid, and Cheminova had to dispose of its waste elsewhere.[42] In addition to reasonings based on necessity, "*Høfde-42-depotet*" was an "away" born out of the torment of winter storms and the hazards of toxic waste.

Third Time's a Charm, or the Return of the Away

As I was interviewing Bjarne Hansen, he asked rhetorically, "How many times can you remove a depot?" Immediately, he answered himself, "They tried two times. Perhaps they will succeed a third."[43]

His observation—intended as a joke—illuminated the inherent paradox of the promise of removal. It was a way of expressing ambivalence toward the forthcoming remediation by drawing on the depot's history. The contradictive statement unraveled how hazardous waste remediation is negotiated through thresholds, risk assessment, and manageability. Using remediation as a metaphor for "gone" evoked an ambivalent sentiment in Bjarne caused by past circumstances, thus making him tell the conundrum-turned-joke.

In 1962, after Cheminova removed the most toxic elements from the site, the depot was reported as closed and secure and remained largely imperceptible for a decade. In 1971, while doing coastal maintenance, the Coastal Department accidentally uncovered and opened up the forgotten depot, to everyone's surprise and, ultimately, dismay.[44] What the depot contained was unknown; years of depositing waste, occasionally in secrecy, and the combination of raw sewage, diverse sorts of residue from pesticide production, mercury, and solid waste, including barrels and debris, made estimations impossible.[45] Negotiations followed, and Cheminova had to remove 1260 m^3 of contaminated sand and re-submerge it on its premises. Nevertheless, the company convinced the department that the remaining depot was best left alone despite storms and coastal erosions indicating otherwise.

Consequently, the site was covered again, but this time sealed with concrete, asphalt, and gravel. The re-storing of waste was the first effort in neutralizing the risk posed by the toxic underground landfill. Suggestions on further precautions—such as fencing off the area or putting up cautionary signs—were measured against the potential of a public environmental scandal, while Cheminova insisted on the harmlessness of the remaining waste.[46] Risk communication about the perils of the depot meant steering particular narratives that avoided public controversy or outrage.[47] Missing signs and fences in arrangement with a subsurface chemical landfill gave the impression of an area where the apparent (and visible) risk was either gone or controlled.

In 1976 and 1978, subsequent storms eroded the depot's cover. The exposed surface generated fear in the local community due to the unseen (yet known) terror of exposure to an unspecified amount of toxic waste. Signs of contamination began to appear in wildlife, and visible signs of yellow water and synthetic stench were a stark reminder of a chemical past and potential forthcoming violence.[48] Years of coastal erosion and storms had moved the sea closer to the depot. Facing the North Sea, the depot was exposed, especially during winter when flooding came after storms. The problem was not water charging the chemical depot but storms inundating the peninsula and, on its return, eroding and unrooting land and materials back into the sea, including built infrastructure such as depots.[49] Contrary to Gunnar Andreasen's ideas about the brilliance of the dune/water system for waste management, every storm was a violent disturbance threatening to drag enormous amounts of chemical waste into the sea.

Several storms in the 1970s attracted the attention of biologists, who found the peninsula inundated by more than 200 different synthetic and toxic chemicals; however, no action was taken.[50] Winter storms in 1980 and 1981 made the situation increasingly urgent, and local authorities asked for intervention from the Danish Environmental Protection Agency (DEPA) and the Ministry for Environment, as the national press became interested in the chemical depot.[51] The DEPA and ministry argued for preventive measures, as removing the depot would be too costly because one remediation could entail another and more expensive course of action. Cheminova downplayed the urgency, trivialized any notion of toxicity, and rejected any implication or responsibility, while biologists and local environmentalists insisted on complete redress.[52] Others thought that opening up the depot fully would unleash all the pollution, instead preferring slow and incremental release and hoping the waste would vanish piecemeal and not result in cataclysmic disruption. A main concern was the urgency of the situation, but the urgency-almost-becoming-crisis became an argument for preserving the status quo rather than changing it, as it was reasoned that the greater the risk, the greater the risk of removing said risk. Discussions gravitated around the type of risk, whom it potentially affected, and how to minimize it—all of which were entrenched in other risk-filled reasonings: political, economic, and technical. In this way, managing the depot's environmental threat was held up against other types of risk, which in turn guided the debate.

As interests shaped perceptions, financial arguments dominated the debate until a memo in March 1981 from a sub-department of the DEPA showed that the pollution had increased and spread.[53] This did not refer to "local" pollution, as first assumed; tides and currents transported the pollution up and down the coast, in and out of fjords. The findings alarmed the national fishing industry and made international headlines from Germany to the UK, with bans on importing Danish fish following. The alarm of an entire sector with substantial export value was too severe.[54] What locals, fishermen, environmentalists, and biologists had insisted on for years—namely that the depot was a genuine threat in terms of potentially unleashing an environmental catastrophe—was an insufficient incentive to do anything, however, the potential risk of a decline in exports was the decisive factor. As such, the authorities were not concerned about environmental consequences or health threats; rather, they feared the economic consequences of terrible international publicity.

A remediation plan was decided and conducted in the summer of 1981. The removal totaled 5,600 tons, 1,200 tons of which was chemical waste, albeit it was not removed but redeposited in Herfa-Neurode, a former salt mine established in 1972, in Hessen, Germany. Cheminova's hazardous waste reiterated the decades-old dilemma of finding a home for substances considered too toxic but which must go somewhere. The first away, the "*Høfde-42*" depot, was replaced with a deeper and seemingly safer place in Germany and thus became someone else's problem.[55] However, because of different and unlikely circumstances, such as a lack of containers for the contaminated sand, reportedly 120 tons of chemical waste remained untouched. A report assessed that the leftover chemicals were safe enough and that gradual leakage would dilute and remove them over time. Mercury was a concern, but a curve in the report showed how the toxicants would slowly leach, with an estimate of being gone completely by 2500. However, it was argued in reports that coastal erosion and storms would speed up the process.[56] Ideas about acceptable thresholds of pollution, or, in other words, how much pollution was allowed or deemed tolerable, were based on estimations and timescales beyond human experience but which were easy to accept as a timeline.

Geographer Angeliki Balayannis shows how bureaucracy shapes removal, connecting it to structures of power that render waste visible and invisible to make it manageable. Additionally, these structures hold a discursive power of establishing thresholds, i.e., that pollution is as much about allowing pollution to a certain extent rather than regulating or removing it. Remediation is about establishing thresholds as safety barriers while normalizing the existence of a toxic waste depot placed in an exposed location, thereby rendering it safe and secure.[57] The ontological status of chemical waste in this case changed from a necessity—it had to be placed somewhere—as proposed by Cheminova, to that of protection. However, protection from what exactly only became apparent as the depot's status as an urgent peril emerged, corresponding to more significant economic incentives. The remediation happened not because of environmental hazards but because influential economic sectors, namely the export fishing industry, were affected.

Living with Contamination

Bjarne Hansen's life is entangled with the depot. He grew up next to it, and his father, Aage Hansen, was a central actor in 1981 when local environmentalists helped secure remediation for the first time. Bjarne's mother was lost prematurely to cancer, a common cause of death in the area, which, according to several conclusions of medical investigations, could only be related circumstantially to the pollution. The reports pointed to individual lifestyle choices, or "bad habits," such as smoking and alcohol, as equally probable causes.[58] In this way, Cheminova and the pollution became part of his most intimate history. Naturally, Hansen has strong opinions about Cheminova, pollution, the depot, the authorities, and the techno-scientific discourses surrounding remediation.

Bjarne Hansen continued his father's struggle in 2010, but in reaction to a process that began in 2000. At the brink of the new millennium, the depot, which had suggestively emptied itself gradually according to reports, was not looking or, to be more accurate, smelling right. A strange, rancid odor arose from the coastline in proximity to the depot. The leaching of chemical waste was far worse than expected.[59] Six years of reports and measurements, followed by difficulties about responsibility, ended in 2006 with a sheet pile wall that functioned as a temporary barrier. At the turn of the millennium, Hansen was a local politician sitting on the local city council. After six years of bureaucratic non-action, he became interested in the debates, and in 2010, after exiting politics, he formed a group, the "*Høfde-42 Gruppen*," consisting of new and old environmentalists working for remediation.[60]

Despite his sentiments and activist approach, Bjarne Hansen and his group have nothing against Cheminova, but their relation to the factory is an ambivalent one. The ambivalence emerges when the source of contamination profoundly impacts community members' lives negatively through adverse health issues or terminal diagnoses but at the same time characterizes community prosperity. Your friend might suffer from diseases related to chemical exposure, while your neighbor enjoys financial success. Bjarne and his group walk a thin line where criticism of pollution, but not the polluter, means that environmentalists must weigh every word. Furthermore, if the factory relocates or shuts down, so does any promise of recognition or responsibility. The ambivalence occurs when perpetrators are, to some extent, also the victims, and criticism can generate a double-negative position of villainization and exposure to chemicals—the group focus on removal instead of blaming, reluctant to undertake and sustain damage-centered activism.[61]

Concentrating on structures that generate environmental harm through thresholds, the group deploys controversy by challenging how much chemical waste remains in the depot, as different numbers have been suggested. A recurring estimation is 120 tons, which originates from reports dating back to the beginning of the 1980s, although the numbers have been refuted several times.[62] Nevertheless, this figure is employed as a solid fact to claim certainty and give the impression that the chemical waste is under control, despite the onto-epistemological vagueness of a 70-year-old chemical waste depot situated beneath the sandy surface of a coastline. Being historically informed, environmentalists witness how numbers, facts, and data forwarded confidently are not as reliable as they appear at first sight. Inundated by discrepancies, the claims surrounding the depot also push another form of ambivalence: how can engineers know what and how much to remediate if it is unclear what the site actually contains?

Six months after announcing the remediation, regional authorities held a citizen assembly and provided the local community with information about the process. Several engineering companies attended as representatives of each cutting-edge technology. Some used water techniques resembling large washing machines to remove toxicity from the sand, and others favored inserting large steel pipes as heating elements to vaporize the toxic waste and then to collect it in what appeared to be large vacuum cleaners or extractor hoods.

Experts and local politicians presented PowerPoints explaining how different technologies could clean up the depot and described the toxic waste's presumed location and any potential difficulties and obstacles. The general point was that the money was secured and the technology was available, thus making it purely a matter of getting started. After the presentation, Bjarne Hansen stood up, and a lengthy exchange followed. Hansen challenged one politician's promise of removing 98% of the depot by asking about waste outside the site. Conventional portrayals of a depository as a confined space, and thus a secure area, were destabilized through questioning boundaries, not only horizontally but vertically, such as how deep the toxic waste had seeped and if the groundwater was safe.

Furthermore, the installed sheet pile wall provided the depot with some stability, security, and duration and was estimated to last for ten to fifteen years. At the time of the meeting, the initial expiration dates had been exceeded as it was installed in 2006. Bjarne Hansen was focused on this issue and raised heavy criticism that included previous experiences with remediations where former promises had been given but not kept. The group challenged dominant epistemic regimes by drawing from historical experiences of previously unsuccessful solutions, and ambivalent attitudes emerged as past promises of removals appeared in the present as a retroactive failure. Still subjected to the effects of buried chemical waste, despite being assured on two occasions that any danger had passed, the environmentalist group nevertheless contested techno-scientific claims[63] by documenting the process through photographs, archival research, and vernacular science, i.e., the art of observing, sensing, and noticing, cultivated from several generations inhabiting the area (see Figure 6.2).

The facts and numbers did not add up and caused a stir among the environmentalists, as removal seemed a matter of perspective. To the authorities, remediation meant removing the waste so pollution would fall below thresholds. As such, removing 98% of the dumped waste meant removing 98% of the perceived risk. However, waste and impact measurements relied on what Michelle Murphy terms "epistemic habits," where the chemical pollution and its impact were artificially compartmentalized through levels and limits.[64] The separation meant that the chemical pollution was depicted as manageable. As Bjarne and the environmentalist group insisted, the static image of a danger happening slowly and incrementally gave associations of latency and thus provoked images of harmlessness. These depictions were produced by the industry, including Cheminova, to compartmentalize—and therefore make more manageable—pollution measurements.[65] As such, there were no higher or lower barriers assigned to the accumulated discharge at the depot, only regulations for each substance, thereby disregarding circulation, accumulation, and cocktail effects. The depot and what it held inside were about perceptibility, but the environmentalists contested traditional claims of certainty by problematizing the numbers upon which they made claims. As much as engineers and authorities tried to simplify the process, the environmentalists tried to make it more complex.[66]

After the meeting, Bjarne Hansen muttered, "I've seen it all before," before adding, "When large sums are involved, the vultures appear. It's not about cleaning or removing the pollution but keeping the wheels turning."

Hansen was referring to a moment when one presenter stated it was "About getting as much remediation as possible for their money." To him, some actors in the remediation process had become businesses dependent on pollution, and over

Figure 6.2 An environmentalist (not Bjarne) documenting on-site soil contamination tests.

Source: Copyright Sebastian Lundsteen.

the years, numerous investigations and measurements had created an economy with little interest in remediation. Hansen had found himself ambivalent in this regard, as the proposed remediation was not what he had imagined.

Remediation, according to the environmentalists, involved exterminating, not controlling, risk—a strategy that had proved unproductive for decades. Acknowledging that there was no such thing as "gone," arguments surrounding techno-scientific reasonings nevertheless promised to remove the pollution.

As Alexis Shotwell argues, purity or cleanliness is not only impossible but also intimately connected to racist and colonial ideologies and is a "bad, but common, approach" despite benevolent intentions.[67] Moreover, building on Balayannis, who shows how remediation is a discursive move embedded in bureaucratic promises of an "away," remediation can likewise appear as restorative justice responding to past injustices. The environmentalist group entangled in these political and bureaucratic narratives sometimes articulated a form of nostalgia: a return to what once was, which still is the aim, despite how unattainable it might be.

This ambivalent position is used strategically by demanding the impossible which exposes techno-scientific logics – not as an objective science dedicated to removing all of the chemical waste – but entwined with political and economic measures. Demanding the impossible is a strategy that sheds light on the ambivalence of remediation by challenging the functions and political dimensions of thresholds. As Bjarne Hansen asserted, "We want them to remove all the poison; of course, we know it is impossible. However, we need to make such demands. Realistically, we just don't get sick just by living here."[68]

Consequently, the nostalgic utopia of a pristine nature or a complete "gone" is seen by environmentalists as an impossibility because previous experiences demanding the possible have only caused partial remediation.

Conclusion

When Bjarne and his group received the news about finding the money for remediation, their immediate reaction was joy. However, as past experiences informed present considerations, elation was replaced by ambivalence. The environmentalists were not against remediation but very much for it, albeit based on certain conditions. Their ambivalence signaled neither undivided optimism nor pessimism but a cautious approach toward remediation, however urgent it appeared.

The environmentalists' continuous ambivalence is a mode of engagement grounded in their insistence on being critical of processes and the science surrounding them. The group has years of lived experience in a contaminated space, seeing repeated pollution and promises of a solution. Moreover, the experience of living with pollution fosters ambivalence that counters types of information that rely on static and simplistic understandings of the perils of chemical waste. Additionally, their ambivalence toward the process has not led to apathy but action, as it reveals how pollution is governed, produced, and maintained through thresholds and how remediation is a part of the same process. Their ambiguous states are not something to be ignored, solved, or overcome but instead offer important insights into a community living in—and with—pollution.

I have attempted to show how ambivalence, as a starting point for the environmental humanities, can attune us to the interwoven nature of the processual, the structural, and the personal, as well as its historical rootedness. Recalling Donna Houston, storytelling avoids causal relations but consists of various "issues and responses to environmental injustice," including emotions, history, and ontological differences.[69] One response to an environmental injustice is ambivalence in relation to the injustice not only after it occurs but also *as* it occurs by taking in historic and ontological differences. As I have tried to show, ambivalence is a mode of expression that contrasts and exposes the inherent paradoxes associated with green transitions (among other things). Furthermore, it walks alongside trouble by following it, being in its proximity, criticizing it, and engaging with it.

Defining Bjarne as a "pain in the neck" simplifies the often multiple and complex stakes individuals or groups experience during urgent transformation. Environmental humanities provide important contributions to societal debates on ecological transformations by intervening in binary renderings that too easily gloss over affective and contradictory attitudes. Approaching environmental governance from other angles, through ambivalence, for example, implies an attunement to emotional attitudes and urges us to pay attention to the multiple, the paradoxical, and the difficult. In doing so, it opens up other avenues of inquiry by asking how ambivalence is produced and what it produces.

Acknowledgments

This chapter is dedicated to Bjarne Hansen for inviting me in, letting me stay in his house, and providing me with food, knowledge, and many hours of conversation. Tak, Bjarne.

Notes

1 Morten Løvig Nielsen, "Se de rørende billeder: Her får miljøforkæmper glædeligt nyt om Høfde 42," *TV Midtvest*, December 6, 2020, https://t1p.de/brrz0.
2 Rune Stubager and Kasper Møller Hansen, *Klimavalget: Folketingsvalget 2019* (Copenhagen: Djøf Forlag, 2021), 1.
3 A "høfde"; *English:* groin. See image for details.
4 "Region Midtjylland," Kemikaliedepotet ved Høfde 42, accessed February 3, 2021, https://t1p.de/grhcp.
5 Marie Linddahl, "Bjarne overtog sin fars kamp mod giftdepot," *TV Midtvest*, January 5, 2021, https://t1p.de/ncgsb.
6 Recent research in (environmental) humanities argues that stories—and storytelling—are both theory and a methodology, in that "storytelling embraces the complexities and contradictions of lived experience, generating grammars at once prosaic and revelatory, textured by everyday realities." I consider this chapter a story that becomes a space of possibilities and "theorizing the structural from embodied experience." See: Pavithra Vasudevan et al., "Storytelling Earth and Body," *Annals of the American Association of Geographers* 113, no. 7 (November 2022): 1–17, doi:10.1080/24694452.2022.2139658.
7 Sebastian Engelberth Hansen, "Miljøforkæmperen Bjarne gentager sin fars succes: Får ridderkors for sin kamp for oprensning af Høfde 42," *TV Midtvest*, November 21, 2021, https://t1p.de/t17ry.
8 "'Rav-Aage' Takkede for Ridderkors," *Dagbladet Holstebro-Struer-Herning*, December 9, 1995.
9 Michelle Murphy, "Alterlife and Decolonial Chemical Relations," *Cultural Anthropology* 32, no. 4 (November 2017): 495, doi:10.14506/ca32.4.02.

10 Panu Pihkala, "Toward a Taxonomy of Climate Emotions," *Frontiers in Climate* 3 (January 2022): 1–22, doi:10.3389/fclim.2021.738154; Jochen Kleres and Åsa Wettergren, "Fear, Hope, Anger, and Guilt in Climate Activism," *Social Movement Studies* 16, no. 5 (September 2017): 507–519, doi:10.1080/14742837.2017.1344546.

11 Max Liboiron, Manuel Tironi, and Nerea Calvillo, "Toxic Politics: Acting in a Permanently Polluted World," *Social Studies of Science* 48, no. 3 (June 2018): 331–349, doi:10.1177/0306312718783087.

12 Klaus Jonas, Philip Broemer, and Michael Diehl, "Attitudinal Ambivalence," *European Review of Social Psychology* 11, no. 1 (2000): 37–39; Sigmund Freud, *Three Essays on the Theory of Sexuality*, trans. James Strachey (New York: Basic, 2000), 65; Sigmund Freud, *Totem and Taboo: Some Points of Agreement Between the Mental Lives of Savages and Neurotics* (1950; reis., London: Routledge, 1999), 29–30.

13 Hili Razinsky, *Ambivalence: A Philosophical Exploration* (London and New York: Rowman & Littlefield International, 2017), 6–9.

14 Ciara Kierans and Kirsten Bell, "Cultivating Ambivalence: Some Methodological Considerations for Anthropology," *HAU: Journal of Ethnographic Theory* 7, no. 2 (September 2017): 26–27, doi:10.14318/hau7.2.006.

15 Kierans and Bell, "Cultivating Ambivalence," 37.

16 Robert D. Bullard, *Dumping in Dixie: Race, Class, and Environmental Quality* (Boulder: Westview Press, 2000), 3; David N. Pellow, *What Is Critical Environmental Justice?* (Cambridge and Medford, MA: Polity Press, 2018).

17 Although climate and environmental disasters are ubiquitous and pervasive, they are not shared equally. See Simone M. Müller, "Toxic Commons: Toxic Global Inequality in the Age of the Anthropocene," *Environmental History* 26, no. 3 (July 2021): 444–450, doi:10.1093/envhis/emab014.006; David N. Pellow, *Resisting Global Toxics: Transnational Movements for Environmental Justice* (Cambridge: MIT Press, 2007).

18 Deana Jovanović, "Ambivalence and the Study of Contradictions," *HAU: Journal of Ethnographic Theory* 6, no. 3 (December 2016): 1–6, doi:10.14318/hau6.3.002; Marien González-Hidalgo, "The Ambivalent Political Work of Emotions in the Defence of Territory, Life and the Commons," *Environment and Planning E: Nature and Space* 4, no. 4 (December 2021): 1291–1312, doi:10.1177/2514848620961737.

19 Julie Sze, *Environmental Justice in a Moment of Danger* (Oakland: University of California Press, 2020), 4.

20 Val Plumwood, *Environmental Culture: The Ecological Crisis of Reason* (London: Routledge, 2007).

21 Donna Houston, "Environmental Justice Storytelling: Angels and Isotopes at Yucca Mountain, Nevada," *Antipode* 45, no. 2 (March 2013): 419, doi:10.1111/j.1467-8330.2012.01006.x.

22 Donna J. Haraway, *Staying with the Trouble: Making Kin in the Chthulucene*, Experimental Futures: Technological Lives, Scientific Arts, Anthropological Voices (Durham: Duke University Press, 2016), 12.

23 Donna J. Haraway, "Situated Knowledges: The Science Question in Feminism and the Privilege of Partial Perspective," *Feminist Studies* 14, no. 3 (1988): 575–599, doi:10.2307/3178066; Haraway, *Staying with the Trouble*, 12.

24 Jørgen Burchardt, "Cheminova—En Generationforurenerens Første År," *Gladsaxe Lokalhistoriske Forenings Årbog 2022* (2022).

25 Jens Engberg, *Det heles vel: forureningsbekæmpelsen i Danmark fra loven om sundhedvsvedtægter i 1850'erne til miljøloven 1974* (Copenhagen: Miljøkontrollen i Københavns Kommune, 1999).

26 "Cheminova Flytter Fra Maalev Til Landets Mest Øde Omraade." *Roskilde Dagblad*, July 1, 1952; Court Transcript of Negotiation Protocol, Landvæsensnævnet for Skodborg-Vandfuld og Hjerm-Ginding herreders 1. nævnskreds, October 15, 1953, 20–24, Lokalarkivet for Thyborøn-Harboøre-Engbjerg.

27 By "Transsubstantiation," I draw on Pierre Bourdieu's conceptualization of the flows between different forms of capitals described as: (…) those forms of exchange which ensure the *transubstantiation* whereby the most material types of capital – those which are economic in the restricted sense – can present themselves in the immaterial form of cultural capital or social capital and vice versa. Pierre Bourdieu, "The Forms of Capital," in *Handbook of Theory and Research for the Sociology of Education*, ed. John G. Richardson (Westport, CT: Greenwood, 1986), 242.

28 Miljøministeriet, *Miljøministeriets orientering om Cheminova: myndighedernes sagsbehandling* (Copenhagen: Miljøministeriet, 1981), 7–11.

29 Miljøministeriet, *Miljøministeriets orientering om Cheminova*, 38–43.

30 Court Transcript of Negotiation Protocol.

31 Finn Bro-Rasmussen and Kirsten Warnøe, "Harboør Tange, Denmark—A Coastal Area Polluted with Toxic Organic Chemicals and Mercury," in *Ecoaccidents*, ed. John Cairns (New York: Plenum Press, 1985), 3–24.

32 "Harboøre-Værket," *Berlingske Tidende*, October 1953.

33 "Rønland-Beboere Klager Til Politiet over Cheminova," *Ringkøbing Amts Dagblad*, August 1956; Miljøministeriet, *Miljøministeriets orientering om Cheminova*, 42–43.

34 Max Liboiron, *Pollution Is Colonialism* (Durham: Duke University Press, 2021), 4–5.

35 Elsebeth Lynge et al., "Cancer Incidence in Thyborøn-Harboøre, Denmark: A Cohort Study from an Industrially Contaminated Site," *Scientific Reports* 11, no. 1 (December 2021): 13006, doi:10.1038/s41598-021-92446-y.

36 "Anmodning om Deponering af Fast Affald i Klitgryden ved Høfde 42," Cheminova, July, 1956, 21–22, Lokalarkivet for Thyborøn-Harboøre-Engbjerg.

37 Miljøministeriet, *Miljøministeriets orientering om Cheminova*, 81–82.

38 Gunnar Andreasen, *Første halvleg* (Harboøre: Forlaget Vest, 1983), 99.

39 Miljøministeriet, *Miljøministeriets orientering om Cheminova: myndighedernes sagsbehandling bilag 2* (Copenhagen: Miljøministeriet, 1981): 103–110.

40 "Den nye Cheminova-fabrik møder nye vanskeligheder: Myndighederne stiller nye krav—Flytter man til udlandet?" *Demokraten,* August 16, 1953; Cheminova, "Anmodning om Deponering af Fast Affald i Klitgryden ved Høfde 42."; Report on inspection of Cheminova, June 26, 1956, 20–24, Lokalarkivet for Thyborøn-Harboøre-Engbjerg.

41 Miljøministeriet, *Miljøministeriets orientering om Cheminova bilag 2*, 107–108; Jennifer Gabrys, "Sink: The Dirt of Systems," *Environment and Planning D: Society and Space* 27, no. 4 (August 2009): 666–681.

42 Miljøministeriet, *Miljøministeriets orientering om Cheminova bilag 2*, 112.

43 Bjarne Hansen, Interview, Oral (September 2022).

44 Nielsen, *Cheminova—en giftig affære*, 81.

45 Nielsen, *Cheminova—en giftig affære*, 29–30; Miljøministeriet, *Miljøministeriets orientering om Cheminova bilag 2*, 126.

46 Miljøministeriet, *Miljøministeriets orientering om Cheminova bilag 2*, 126–127.

47 Alex Zahara has a similar point, see: Alex Zahara, "On Sovereignty, Deficits, and Dump Fires," in *Inevitably Toxic: Historical Perspectives on Contamination, Exposure, and Expertise*, eds. Brinda Sarathy, Vivien Hamilton and Jannet Farrell Brodie (Pittsburgh: University of Pittsburgh Press, 2018): 261.

48 Miljøministeriet, *Miljøministeriets orientering om Cheminova*, 161; Øjvind Kyrø, "Havet sletter ikke alle spor," *Berlingske Aftensavis Weekend edition*, Autumn 1981: 5.

49 Cheminova-Gruppen, "Giftdepotet ved høfde 42 ... fjernet i 11. time: en status over fjernelsen af et af Cheminovas giftdepoter på Harboør Tange," *Cheminova-gruppen* (May 1983): 2–4.

50 Gudrun Wium-Andersen and Søren Wium-Andersen, *Cheminova-Sagen 1978: En Miljøsag Belyst ved Akter og Avisdebat* (Ørsted: Nucleus, 1979).

51 Miljøministeriet, *Miljøministeriets orientering om Cheminova bilag 2*, 236–245; Nielsen, *Cheminova—en giftig affære*, 88–89.

52 Karl Martin Pedersen, "Cheminovas depot en varm kartoffel, men risikoen for en katastrofe er ikke akut," *Ringkjøbing Amts Dagblad* (September 1978); Cheminova. *Argumentation against removing depot by Cheminova.* Letter. Århus: Miljøstyrelsen Århus, May 1981. Located in Miljøstyrelsen, *Høfde 42 (Cheminova) (1981–2003),* 218: Undersøgelser, Harboøre Tange 1981–1982 m.m. Archival ID, 931 145.

53 Wium-Andersen and Wium-Andersen, *Cheminova-Sagen,* 1978.

54 Kyrø, "Havet sletter ikke alle spor," 12.

55 The mine is now a contested place, as local communities fear for the environmental repercussions of living next door to one of Europe's largest toxic depots, causing ambivalence elsewhere. As such, "Away" does not exist. Ludger Fittkau, "Herfa-Neurode in Hessen—Die größte unterirdische Giftmüll-Deponie der Welt," *Deutschlandfunk,* April 4, 2019, https://t1p.de/m48wm,

56 Miljøstyrelsen and Cowiconsult, "Fjernelse Af Kemikalieaffaldsdepot Ved Høfde 42 På Harboøre Tange. Afsluttende Undersøgelser," September 1981, 6.

57 Angeliki Balayannis, "Toxic Sights: The Spectacle of Hazardous Waste Removal," *Environment and Planning D: Society and Space* 38, no. 4 (August 2020): 772–790, doi:10.1177/0263775819900197; Michelle Murphy, "Chemical Infrastructures of the St. Clair River," in *Toxicants, Health and Regulation since 1945*, eds. Soraya Boudia and Nathalia Jas (London: Pickering and Chatto, 2013), 103–116; Max Liboiron and Josh Lepawsky, *Discard Studies: Wasting, Systems, and Power* (Cambridge: The MIT Press, 2022).

58 Lars Kamstrup, "Stor kræftstigning i Thyborøn-Harboøre ikke unormalt," *Lemvig Folkeblad*, April 27, 1988; Lynge et al., "Cancer Incidence in Thyborøn-Harboøre, Denmark."

59 Hanne Pedersen, "Notat om Forureningsproblematikken ved Høfde 42, Cheminova-Hullet og Området ved Spildevandsledningsbruddet," March 2001.

60 In Danish: Remove the Poison.

61 Eve Tuck argues against "damage-based research" that focuses on the harm communities suffer because of structural (environmental) discrimination. Tuck suggests recentering research based on desires and aspirations. Hansen and the group do not explicitly use the term "damage-based activism." I interpret their strategies as a reflection and a way of engaging with the concept. It should be added that the group occasionally uses "damage narratives" as a strategy. Eve Tuck, "Suspending Damage: A Letter to Communities," *Harvard Educational Review* 79, no. 3 (October 2009): 409–428, doi:10.17763/haer.79.3.n0016675661t3n15.

62 Knud Jakobsen, "Udsivning af gift fra Høfde 42 standses ikke," *Thisted Dagblad*, October 2003.

63 A report from 2021 shows a high incidence of cancer in the population around Harboøre Tange: Lynge et al., "Cancer Incidence in Thyborøn-Harboøre, Denmark."

64 Michelle Murphy shows how pollution and exposure are problematically entangled in epistemic habits, where facts and claims rely on thresholds and information is produced by the chemical industry. Murphy, "Alterlife and Decolonial Chemical Relations."

65 For environmental effects and harm in connection with slowness and delay, see: Rob Nixon, *Slow Violence and the Environmentalism of the Poor* (Cambridge: Harvard University Press, 2013); Murphy, "Chemical Infrastructures of the St. Clair River."

66 Alex Zahara shows how risk depends on a particular type of strategic communication that simplifies information as experts conflate complexity with intelligibility. See: Zahara, "On Sovereignty, Deficits, and Dump Fires."

67 Alexis Shotwell, *Against Purity: Living Ethically in Compromised Times* (Minneapolis: University of Minnesota Press, 2016), 9–16.

68 Interview, Bjarne Hansen.

69 Houston, *Environmental Justice Storytelling*, 432.

7 Global Waste

On the Ambivalence of Wealth, Health, and Contradictory Development Models

Iris Borowy

Introduction

On July 31, 2023, the National Development and Reform Commission (NDRC) of the People's Republic of China announced a series of policies "so as to lift residents' willingness to consume." Li Chunlin, deputy director of the NDRC, listed several measures, including improving income distribution, instituting paid leave, and introducing flexible working hours as measures to increase disposable income—all designed to "spur consumption of a number of products, services and areas, including new energy vehicles (NEVs), home appliances, electronics, catering, culture and tourism, as well as in rural areas."[1] In a narrow sense, the aim was to revive the national economy and boost "high-quality development" after the downturn experienced during the Covid epidemic. However, in a broader sense, it exemplifies the nexus between economic development, on the one hand, and production and consumption, on the other. It may be particularly obvious in view of the spectacular recent development of China, whereby growth in industrial production has propelled the country to the status of a "manufacturing superpower," providing 28.9% of global manufacturing output in 2019, while also reducing the number of those living under a poverty line of $1.90 per day by a staggering 800 million people.[2] In principle, though, these connections hold true everywhere in terms of the economic development of the last two centuries: a rise in income, wealth, and overall living standards has been achieved with—and reflected by—an increase in the production and consumption of things.

Wealth, Health, Production, Consumption, Waste, and Destruction: Connecting the Dots

Things are good for human well-being. Until they are not. They are coveted by governments, corporations, and citizens alike because they form the basis of improved lives and perceived private, corporate, and (in the case of GDP) governmental success. Improving living conditions for low-income societies is difficult to imagine without making more things available: more buildings for housing and services, more means of transportation for people and goods, and more consumer items to make life easier, more pleasant, and more interesting. The production of

DOI: 10.4324/9781032627984-10

these things provides wages and salaries for workers, profits for companies, and taxes for governments, all of which help finance safe, healthy, and attractive private living conditions as well as public infrastructures for education and healthcare.

But at some point in time, sooner or later, these things are discarded because they no longer work, their purpose has been fulfilled, they are replaced by newer, better or more fashionable and hipper things, or because their owners realize they were bought erroneously in the first place. They become waste. The more things there are, the more things will into waste, and they do so increasingly quickly. As such, they join the large quantities of materials that became waste even before they existed, i.e., the waste generated during the direct and indirect production processes necessary to create houses, machinery, and consumer goods, as well as the energy necessary to produce and apply them: extraction overburden, waste rock, tailings, contaminated mud and sandy materials left next to where minerals were mined or behind tailings dams, which can break, flood communities, and contaminate regional ecologies, or leach acute or long-term toxicants.[3] Waste is not the only aspect revealing the ambivalent nature of industrial developments, but it may be its most direct manifestation, the most tangible connection between the good and the bad of modern lives, "the canary in the coal mine. It shows how much of an impact we're having globally, as a species, on the planet as a whole."[4]

The scales are immense. Figuratively and realistically speaking, coal mines have driven the development of unprecedented wealth and well-being. According to the Maddison Project, global GDP per capita rose from $1,174 in 1820 to $2,012 in 1900 and $14,700 in 2016 (all expressed in 2011 US dollars), a more than twelve-fold increase over two centuries despite a simultaneous seven-fold increase in the global population—a spectacular rise without precedent in human history.[5] While these increases have been uneven, masking substantial within- and between-country inequalities, they have involved close to the entire world population. While in 1800, 76% of the people then living earned less than two dollars per day, by 2022, it was a mere 8%.[6] These economic developments have had tangible effects on population health. While wealth is not a sufficient determinant of health, it nevertheless provides the means for improved living standards, public health policies, and medical research, all of which improve health in the long run. So, though important differences between continents remain, virtually everywhere people's life expectancy has doubled over the last century.[7] Health improvements are also expressed in body height, which reflects nutrition and disease burden experienced during infancy and adolescence. Bodily height has increased unevenly—but in all countries in the world—over the last century, indicating that fewer children have gone hungry and suffered serious diseases than in earlier generations, resulting in positive effects throughout their lives.[8] Economic historians Robert Fogel and Dora Costa have coined the expression "technophysio evolution" to describe a virtuous circle of better nutrition, better health and work capacity (physically and cognitively), more and more innovative work output, higher incomes and living standards, more investments in the next generation (broadly conceived, including health in utero as well as child education), and better nutrition and health.[9] High incomes and wealth can make people not only healthier, but also happier. In a recent global ranking of happiness, "GDP per capita is 20 times higher in the top ten than in the bottom ten countries."[10] Per capita GDP is also correlated negatively with infant mortality rates and maternal deaths, and positively with life expectancy, years of schooling, learning outcomes, working hours, and overall life satisfaction.[11] In other words, all else being equal, people in rich societies tend to live longer, suffer fewer deaths in infancy and during pregnancy, are more educated, work less, and are happier than those in poorer societies.

They also produce more waste, which regularly increases in line with rising incomes.[12] The scale is staggering. At present, the "technosphere" defined as all human-made products and materials currently in use or usable (including buildings, roads, energy supply structures, consumer goods, farmlands, managed forests, etc.) may amount to 30 trillion tons (Tt), equivalent to more than 50 kg per square meter of the Earth's surface.[13] Only a small part is presently reabsorbed into the material cycle after use. The vast majority is just waiting to become waste. Even if the global economy stopped growing today, the waste emerging from human societies would be high and continue to expand for decades.

The component that is most visible and most directly connected to consumer goods is household waste, usually labelled municipal solid waste (MSW). At the time of writing, people around the world generate 2.24 billion tons of MSW per year, a number that is expected to increase to 3.88 billion by 2050.[14] Amounts vary widely. High-income countries are home to 16% of the world's population but produce 34% of global waste. Geographically, at the moment, the East Asia and Pacific region, i.e., an area with a large population living in wealthy or emerging economies, is generating most waste overall, though not per person. However, as living standards increase in countries still suffering widespread poverty, their societies eventually "catch up," in which case waste production is projected to rise dramatically in South Asia and Sub-Saharan Africa.[15]

Discarded items consist of a broad range of materials, only the most important of which will be briefly named herein. Between one-third and over half of all waste is organic in nature, including food waste, which is estimated to represent roughly one-third of all food produced, literally wasting land, water, energy, and labor that went into its production.[16] Another large and rapidly growing material group is plastics, which barely existed until the Second World War, only to experience a spectacular rise in production thereafter, thus reflecting its versatility and low cost. One study estimated that, as

> of 2015, approximately 6,300 MT of plastic had been generated, around 9% of which had been recycled, 12% was incinerated, and 79% accumulated in landfills or anywhere in the environment, including in the oceans. If current production and waste management trends continue, roughly 12,000 Mt of plastic waste will be in landfills or in the natural environment by 2050.[17]

A smaller but even faster-growing waste stream is composed of e-waste, a particularly problematic conglomerate of materials that are difficult and expensive to discard without releasing toxicants into the environment. A 2022 report by the

UN Institute for Training and Research (unitar) estimated that the world generated 53.6 MT of e-waste (an average of 7.3 kg per person that year), 5.1 Mt, or 9.5%, of which were exported, partly in uncontrolled, illegal forms. If trends continue unabated, e-waste is on track to reach 74.7 Mt in 2030 and a stunning 110 Mt in 2050.[18] These amounts of MSW are dwarfed by other forms of waste. Globally, societies produce 12.73 kilograms of industrial waste per capita per day, almost 18 times that of MSW. Respective numbers are 3.35 kilograms of agricultural waste and 1.68 kilograms of demolition waste. Less important in quantity but significant in impact are 0.32 kilograms of hazardous waste of various origins.[19]

These waste types reflect the wealth and convenience of modern societies: both organic and agricultural waste originate from the large farming and food production operations that feed an ever-growing number of people, with famine disappearing from entire continents where it used to be commonplace. Industrial waste is the outcome of producing the things that surround our lives, ranging from cars, to cigarettes, to medicine. Plastics reflect lives filled with colorful toys, affordable household items, throw-away bags, hygienic single-use medical equipment, insulation for safe electricity, HIV-preventing condoms, plastic bottles, and countless other items, ranging from the life-saving to the silly and superfluous. E-waste is the result of the innumerable machines that help an ever-growing number of people keep their homes warm or cool and their food fresh. They help people cook their food, wash their clothes, watch the news, write research papers, and stay in contact with loved ones.[20] Many lives, especially of women around the world, have been transformed immeasurably by electronic machines that have helped reduce the time needed to do household chores traditionally expected of them. But collectively, these waste streams also form a huge and growing burden on environmental and human well-being.

Waste gets thrown away, and the sprawling landfills around the world are a stark reminder that "away" is a place, and since more than half of global waste ends up in landfills or on open dumps, these places are spreading.[21] Landfills are inherently problematic. They emitted an estimated 1.6 billion tons of CO_2 equivalent greenhouse gas emissions in 2016, anticipated to increase to 2.6 billion tons per year by 2050. Overall, greenhouse gas emissions emanating from solid waste treatment and disposal amount to approximately 5% of global emissions.[22] In 2014, the 50 largest active dumpsites affected the daily lives of 64 million people, with those living nearby showing elevated prevalence of diseases such as diarrhea, headaches, chest pains, irritation of the skin, nose and eye issues, typhoid, or stomach ulcers.[23] Furthermore, these "monster dumps" are only the most visible manifestations of waste disposal. Around the world, waste is stored, dumped, and burnt in myriad ways, and every form being problematic, in that it pollutes soil (when landfilled), water (when dumped in waterways), and air (when incinerated). Contaminants are absorbed by plants in affected areas and then by animals and humans that eat them, creating as yet little understood health burdens. In Europe alone, 342,000 contaminated sites were identified in 2014, corresponding to 5.7 per 10,000 inhabitants. Though information is often limited, poor hazardous waste management has been connected to several forms of cancer, adverse birth outcomes, and congenital anomalies.[24] In what Rob Nixon famously labeled "slow violence," waste with various degrees of toxicity causes a variety of long-term health burdens and diseases, especially for people who are too poor, too powerless, or too marginalized to protect themselves.[25] Environmentally, the waste sector "contributes significantly to the triple planetary crisis of climate change, biodiversity and nature loss, and pollution."[26] Even unproblematic forms thereof create a social burden, as waste management places financial strains on municipal budgets everywhere, often constituting the largest budgetary items in cities in low-income countries, subsequently diverting funding from other important areas.[27] Finally, municipal waste represents a loss of (often scarce) resources, necessitating production from virgin resources and thereby provoking new environmental burdens from mining waste.

Eliminating, or even reducing, the harm caused by waste is tricky, since it is not easily separated from the benefits with which it is also connected. This applies directly to landfills in low-income countries—places that have been seen as quintessential examples of environmental injustice, where the rich in the Global North dump their toxic waste onto the poor in the Global South.[28] Without glossing over real instances of health hazards and criminal behavior, scholarly attitudes have recently shifted from outrage to more nuanced views that acknowledge a more complex picture of North–South, North–North, South–South, and South–North waste trade. These flows often involve elaborate repair and recycling processes and a lively trade in secondhand products while providing livelihoods for hundreds of thousands—and possibly millions—of people around the world, often working in the informal sector.[29] This is even true for the most notorious cases, such as Agogbloshie in Ghana and Guiyu in Hunan Province, China. Both became known as atrocious dumping grounds for e-waste from industrialized countries in Europe and North America and for poor people trying to scrape a living retrieving raw materials by crushing and burning discarded machines, working without protective gear, and while breathing in toxic fumes. In both cases, more recent research has shown that, by the 2010s, the large majority of waste being treated originated from the region and neighboring countries, and that, rather than simple dumping grounds, these places were hubs of complex trade networks, giving work to thousands of people, including skilled technicians providing qualified repair services.[30] Other studies have pointed to the relative autonomy of informal waste pickers on landfills, leading one researcher to compare landfills to modern-day commons, "under threat not only from a "hygienic enclosure" stimulated by sanitary requirements but also increasingly by new technologies, such as anaerobic digestion and capital-intensive waste-to-energy plants, that have begun to displace landfilling in the Global North."[31] Maneuvering between belittling waste workers as passive victims and romanticizing work on toxic landfills while doing justice to the ambivalence of reality is the challenge faced by scholars of waste.

However, landfills only represent one component of an overall system that connects production, consumption, waste, and income. Admittedly, the situation is multi-factorial because waste as well as physical and mental well-being are influenced by a variety of elements, such as economic inequality, social support systems, power relations, and others. Nevertheless, the core ambivalence remains that waste is connected to structures that poison human and non-human nature as well as to structures of economic wealth that form the basis of good health and long and happy lives—something many people would define as the "good life."

The connective element is mass production and consumption. On the one hand, they drive private, corporate, and public incomes that can be used in ways that improve people's health, happiness, and life satisfaction. On the other hand, they feed the mounting quantities of waste that pollute the soils, water, and air of the world, notably for politically and economically vulnerable communities (see Figure 7.1).

Separating the good life from waste requires the decoupling of one and/or the other from production and consumption. Decoupling economic wealth from waste generation means societies would continue to have high levels of production and consumption while generating little waste. Decoupling economic wealth from human well-being means societies would enjoy similar or higher levels of well-being while consuming fewer products, therefore generating less waste. Both possibilities have been discussed for some years as part of larger economic concepts, namely, the circular economy (CE) and degrowth. Both have their roots in heterodox economic concern about the high-production, high-consumption developmental model that formed in industrialized countries in the 1960s and 1970s and gave rise to alternative economic ideas as well as the environmental movement.[32]

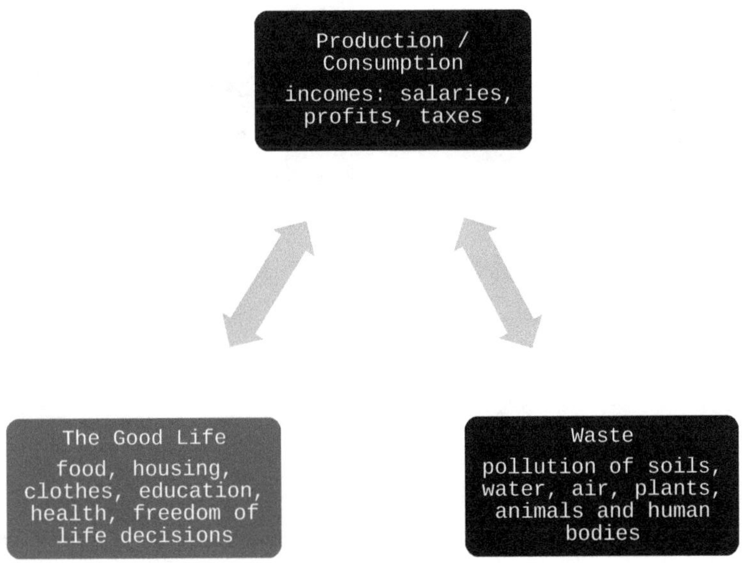

Figure 7.1 The good life-waste nexus in conventional economic development.

Source: Copyright by Iris Borowy.

While both aim at profound socio-economic transformations that will reduce waste, their approaches, rationales, and methods differ substantially. Both have their followers, and both are based on a different set of assumptions about social processes and human nature. Moreover, both leave room for ambivalence.

Decoupling Wealth from Waste Production: The Circular Economy

Conceptually, the circular economy mimics organic material transformations in nature, where plants and animals live, eat, excrete, die, decompose, and provide building materials for new forms of life in a perpetual cycle of material transformations that sustain life on Earth. Ideally, a cycle of human-made products would function in the same way, connecting production, use, disassembly, and recycling in an endless circle of use and reuse. Ideally, it would make all waste disappear by converting it from a final product of a linear production format to a transitory stage within a constant process of economic and material transformations (see Figure 7.2).

Kenneth Boulding's 1966 text "The Economics of the Coming Spaceship Earth" is generally considered a formative work on the idea of circularity, albeit without using the term per se. He invoked a comparison of modern societies with astronauts who lived with finite space and resources, obligated to sustain themselves in a "cyclical ecological system."[33] However, for many years, the concept was not followed up to any great extent, and the first use of the term "circular economy" appears to have been in a book entitled "Economics of Natural Resources and the

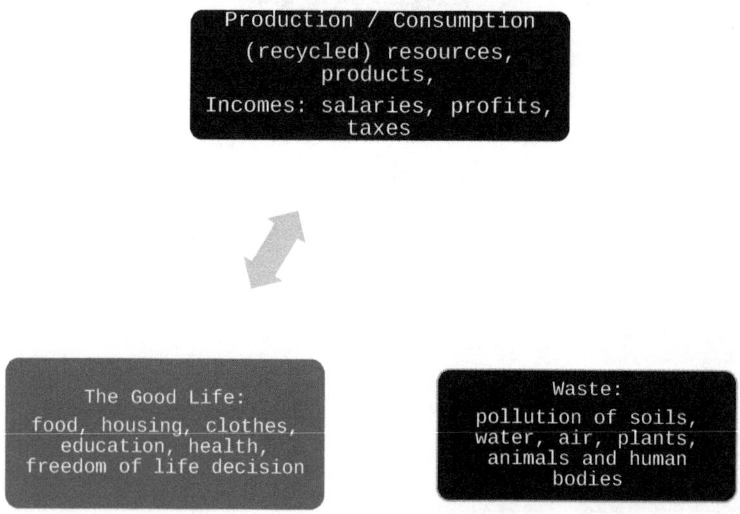

Figure 7.2 The circular economy decoupling of the good life from waste.

Source: Copyright by Iris Borowy.

Environment" by David W. Pearce and R. Kerry Turner, published in 1990. Other landmarks were the publication in 2002 of the book *Cradle to Cradle: Remaking the Way We Make Things* by William A. McDonough and Michael Braungart, as well as the creation in 2010 of the Ellen MacArthur Foundation, whose central mission is to promote the idea.[34]

Inevitably, waste has been at the center of the concept of CE. A common definition flatly declares that CE is "an economic system aimed at eliminating waste and the continual use of resources."[35] However, other definitions are more modest and do not aim at full circularity but merely at various degrees of waste reduction and recycling. The US Environmental Protection Agency, for instance, explains that a "circular economy keeps materials, products, and services in circulation for as long as possible."[36] Similarly, the World Bank explicitly accepts that products will eventually turn into waste and will be discarded, stating that the intention of CE is "to extend the lifespan of consumables and to minimize the environmental impact of final disposal."[37]

This tension between either eliminating or merely reducing waste ties into its contradictory setting at the intersection between ecology and economy. On the one hand, its ecological prospect is enticing. As one author of waste studies flatly states, its ideal form promises no less than that it "could allow us to live within planetary limits."[38] The Ellen MacArthur Foundation echoes these hopes that CE will address key environmental challenges but in more capitalist terms when, on the same page, it appeals to commercial interests, evoking "a resilient system that is good for business, people and the environment."[39] The European Commission promoted its first circular economy action plan as including measures designed "to help stimulate Europe's transition towards a circular economy, boost global competitiveness, foster sustainable economic growth and generate new jobs."[40]

There is a profound ambivalence in the way CE invokes the same growth-centered system that has caused the dramatic growth in global waste in order to eliminate this waste. It also involves the commodification of a solution to the waste problem: private companies, whose business depends on producing products designed to become waste, are called on to increase their business by producing less waste or, if possible, none at all.

No doubt, this unashamed embrace of economic interests has made the system easily acceptable in both national and international administrations and in business circles. In 1991, Japan was the first country to enact a CE law, and among other policies, the government established classes and recycling laboratories in schools, provided enterprises with circular trading markets, introduced customer-friendly collections, and created waste recycling stations. On the level of waste management, the policy was remarkably successful. According to one study, by 2007, Japan was recycling 98% of its metals, recovering 74–89% of materials from electrical equipment, and landfilling merely 5% of its waste. Germany and China followed suit in having CE laws.[41] The German Closed Substance Cycle and Waste Management Act (*Kreislaufwirtschafts- und Abfallgesetz*), adopted in 1996 and revised in 2012 as part of EU harmonization, stipulated that waste prevention should come before re-use, re-use before recycling and other recovery, and recycling before disposal.[42] An updated European version included more stringent regulations restricting single-use items and at-source separation.[43] By 2022, 70% of waste was recycled and 12% used for the generation of electricity.[44] The German law, along with the Danish Eco-Industrial Park of Kalundborg (see below), served to inspire China, with the concept of CE introduced in 2000 and entered into State Council directives in 2005. The concept found its way into the eleventh and twelfth five-year plans, and, in 2008, Chinese authorities promulgated the Circular Economy Promotion Law, which came into force in 2009.[45] Substantial increases in recycling rates have been recorded since.[46] By 2017, the UK, France, Canada, The Netherlands, Sweden, and Finland had all embraced different versions of CE.[47] Most recently, in February 2023, the government of Mexico issued a similar law.[48]

International organizations have also come on board. In December 2015, the European Union adopted an action plan for CE. Three years later, it adopted a set of measures and indicators for monitoring progress toward a CE system and describing actions to strengthen CE for 27 of the most common materials used in the economy.[49] CE has also been endorsed by the UN Centre for Regional Development in Asia and the Pacific, which has held Regional 3R and Circular Economy Forums every year since 2009, as well as a World Circular Economy Forum, which has held meetings since 2017, connecting thousands of participants from political life, academia, and the business sector.[50] In 2021, both the OECD and the G-20 also endorsed a transformation toward circular economies, followed by the World Bank a year later.[51]

However, the widespread embrace of the concept reflects its compatibility with market structures more than its effectiveness in waste prevention. At any rate, it has clearly not led to a noticeable decline in waste generation. As a case in point, China, albeit one of the first countries to adopt the concept, has also experienced a dramatic increase in waste production along with its amazing economic development.[52] Similarly, according to a 2022 World Bank report, "more than 87 percent of EU resource consumption still comes from primary materials, and overall EU waste generation keeps increasing."[53] Often, this may not mean that recycling is failing to grow but that underlying economic growth is even bigger. In his classic book on scrap recycling, Adam Minter points out that between 1960 and 2010, recycling in the US increased from 5.6 million to 65 million tons, which was a huge success. However, during the same period, garbage generation grew from 81.1 million to 249.9 million tons, thus also vastly increasing the amount of waste discarded into landfills or incinerated.[54]

Sometimes, increased recycling rates can be offset by the poor local implementation of CE stipulations by the central government—on top of increased economic activity.[55] But even when energetically translated into official policy, the overall effect can be ambivalent or counterproductive. Thus, when a large-scale industrial park was constructed in Guiyu, China, in order to integrate e-waste recycling into a professional CE system, recycling processes clearly did become more environmentally benign. But the park also displaced many informal, often family-run workshops and appeared to reduce the overall extent of recycling taking place.[56] Oftentimes, CE endorsement seems to express administrative intent (or at least a performance of intent) more than effective policy.

Meanwhile, however, manufacturers are intrigued by the prospect of recycling as a new "global resource frontier."[57] This notion is particularly true for the fast-growing e-waste sector, which contains a large number of environmentally problematic but also economically valuable substances. According to Unitar estimates, only 17% of global e-waste in 2019 was collected and recycled, leaving valuable resources such as gold, silver, copper, and platinum (worth approximately $57 billion) dumped in landfills or incinerated.[58] The successful application of CE principles for e-waste, therefore, promises an irresistible combination of being (and looking) environmentally responsible while increasing profits.[59] And e-waste is only one potential sector in this regard. Estimates of the resource content within waste in general drive visions of huge profits. In 2021, for instance, a corporate study estimated that a CE could generate $4.5 trillion worth of additional economic output by 2030, and it provided a list of companies that had claimed to have profitably integrated CE into their business model.[60] Similar considerations drive municipalities, especially in view of rising prices for rare raw materials such as magnesium, antimony, vanadium, and platinum metals, which are essential for complex modern technologies. Cities thus increasingly turn to urban mining as a solution.[61]

Relying on recycling as a centerpiece of the waste reduction strategy is not unproblematic; it works only if and when recycling is technologically and economically feasible. Economically, the process is demand-driven and labor-intensive, so that turning waste into a resource is more likely to be found in the Global South, where wages are low, than in the North.[62] Besides, there are technical limitations in this regard, since nothing is completely recyclable, inevitably leaving some leftover material, and often recycled products are of lower quality than the original.[63]

Furthermore, while recycling forms a necessary component of CE, it is not the same as a CE system. In principle, CE goes far beyond that, requiring a profound structural transformation, but there is no clear, universally agreed understanding of such a transformation. The lack of such a clear definition can have contradictory effects, on the one hand, standing in the way of effective economic restructuring but, on the other hand, facilitating small-scale changes. At present, CE theory propagates five business models, which, to varying degrees, exist in some shape or form:

1 Using circular inputs: choosing inputs from renewable, bio-based, regenerative, or recycled sources for both energy and material resources.
2 Produce use extension: entailing an expansion of the use and the provision of repairing, refurbishing, and re-manufacturing, which will prolong the use of a product before disposal and, thereby, reduce the overall quantity of waste.
3 Product as a service: a shift from selling a product to consumers to selling the service this product is meant to provide.

4 Sharing platforms: involving sharing products by offering services to many users—a model that has grown rapidly in recent years through initiatives such as Airbnb, mobike, Uber, etc.
5 Resource recovery: this method addresses the end of a product's use and when it is discarded, but value is drawn from resource recovery through recycling of the material or through waste-to-energy recovery (incineration). This model may involve innovative connections between different manufacturers who enter into a symbiotic relationship of mutual use of each other's waste as a resource.[64]

These models reveal the spectrum of arrangements considered CE, none of which is circular in the literal sense of the word. Nevertheless, apart from the first, which has a minimal impact on production and consumption, leaving intact all conventional elements of a linear economy, all the others involve different degrees of economic transformation. Business activities such as planning, pricing, and marketing all follow a different logic if the aim is to gain revenue from service rather than from products. Profits would no longer be increased by designing products for quick disposal, thereby necessitating new purchases; instead, they would be increased by providing sturdy, long-lived designs. Such designs could benefit from making use of secondary markets and combining scrap material from various production lines. Crucially, such new economic formats would also require—and supposedly lead to—psychological and cultural transformation, ranging from the habits, expectations, and attitudes of consumers to new concepts of ownership.[65] While businesses are central to implementing such new models, administrations can play an important role in driving a shift to such models through policies such as extended producer responsibility, standardization, regulations for sustainable public procurement, and science-based targets.[66]

The record of real-life implementation is ambivalent. One strand has been a burgeoning sharing economy, in the form of companies such as Uber, Airbnb, or mobike. Originally hailed as a way to provide services outside of the formal economy, their rapid commercialization has revealed a problematic side to these initiatives. Airbnb, for instance, may have increased, rather than decreased, resource consumption, as cheap prices have triggered a rebound effect and increased tourism. In addition, by generating higher income than long-term rent, it has also withdrawn living space from urban areas, particularly for low-income citizens unable to pay prices in line with short-term tourist stays.[67] As a positive point, however, such sharing schemes have demonstrated the degree to which transformations in economic planning and consumer behavior can occur quickly.

Similar ambivalent findings apply to corporations, a growing number of which have begun integrating CE elements into their programs and publicizing it to improve their brand. As one example, H&M has launched a program to collect, recirculate, and recycles clothes bought at their stores, while Adidas is marketing a "made to be remade" shoe designed for easy recycling.[68] They are in good company. A 2019 analysis of sustainability reports from 123 companies working in land-use-intensive sectors in five continents found an overwhelming endorsement of CE with only scant use of alternative sustainability concepts.[69] The outdoor provider, Patagonia, may have gone furthest in this regard, or at least has been most vocal about its efforts. The company describes taking inspiration from the cradle-to-cradle book and beginning an uphill struggle to integrate CE methods into their business. After first attempts failed when only few clothes were returned, subsequent attempts became more successful when Patagonia adapted entire product lines to a circular system and when customers became more familiar with discussions about the environmental impact of clothing. At present, the Patagonia website boasts its ambition for t-shirts, namely, "to own all its waste across the lifespan of this tee, which is made from a blend of preconsumer mechanically recycled cotton and postconsumer chemically recycled cotton that creates like-new virgin fiber quality—and might never see a landfill."[70]

Another strand consists of hubs designed to pair complementary industries and services. One often-cited example is the eco-industrial parks of Kalundborg, Denmark, whose various schemes dedicated to reusing waste products have created an intricate web of corresponding units. Over the course of a few years, the power plant has provided heating for residential homes as well as given cooling water to an on-site fish farm, steam to a refinery, gypsum to a wallboard maker, and fly ash to road building and cement production. In the process, it has drastically reduced its generation of fly ash, scrubber sludge, sulfur, sulfur dioxide, and carbon dioxide. Eco-industrial parks in other countries have sought to emulate this development, though the concept is still something of a niche phenomenon.[71]

These economic benefits and ready suitability for an existing market economy have doubtlessly spurred the spread of the concept and appear to have resulted in real environmental gains. This can be seen as a strength, in that many initiatives can begin on a local level; companies, citizens, and/or local administrations can engage in independent projects and thereby make use of numerous people's ideas, creativity, and ingenuity. The focus is on practical solutions, eliminating any need for political discussions on ideology and power. This decentralized approach has made it possible to produce a track record that can provide inspiration and support a belief in feasibility, while profit serves as a powerful incentive. However, this modest approach comes at a price.

CE, due to its easy compatibility with existing linear economic rationales, lends itself to greenwashing, but even if all actors are transparent and honest in their policies, the unambitious implementation prevents far-reaching effects, or, as Savini commented, "Current policy packages are a far cry from the radical principles of the circular economy's earliest applications."[72] In its present form, which relies heavily on recycling, waste—paradoxically—becomes a solution to the problem of overconsumption rather than its manifestation. Thereby, the system justifies "a vicious co-dependency, in which waste accumulation and material/energy supply bypasses the more fundamental source of ecological problems: ever-growing consumer capitalism."[73]

In that sense, CE combines the theory of the perfect elimination of waste with the practice of upgrading it to an essential building block of the economy. It also combines the benefits of an a-political approach that facilitates easy cooperation of all stakeholders who focus on practical solutions with the weakness of an a-political approach that sidelines discussions about politics, ideology, and power. Arguably, such discussions are necessary to get to the heart of a transition that is commensurate with the extent of the challenge.

A far more political and ideational approach to the systemic malfunctioning of the economy is degrowth.

Decoupling the Good Life from Production and Consumption: Degrowth

Degrowth is both a movement and a set of ideas regarding the perceived need for a profound transformation of socio-economic structures. It has its modern origins largely in the same criticism that emerged in industrialized countries on environmental and social developments during the 1970s, which also energized the push toward CE thinking. Both movements count the 1966 publication by Kenneth Boulding, cited above, as well as the 1972 best-selling study "Limits to Growth," as part of its intellectual pedigree.[74] Other formative publications include Nicholas Georgescu-Roegen's complex study on entropy, which uses the laws of thermodynamics to explain that continued economic growth is impossible, and Herman Daly's work, which defines the economy as a means to an end, positioning it in the middle ground between the physical world (ultimate means) and spiritual aims (ultimate needs).[75] The first person to use the expression "décroissance" as a positive goal was French intellectual André Gorz in a debate in 1972. The movement of a radical critique of a growth-based economy then gained momentum with a UNESCO conference in Paris, entitled "Défaire le développement, refaire le monde" (Undoing development, redoing the world), attended by over 800 environmental activists and postdevelopment researchers. It was followed by a series of conferences specifically dedicated to the idea of degrowth, beginning with meetings in Paris (2008), Barcelona (2010), Montreal and Berlin (2011), Venice (2012), and Leipzig (2014).[76]

While emerging from similar analyses of what was wrong with existing developments, the degrowth movement drew very different conclusions. Waste generation is not considered the central problem but rather a symptom of a much larger problem. The Research and Degrowth website, which acts as a centerpoint for relevant findings, defines degrowth as "a downscaling of production and consumption that increases human well-being and enhances ecological conditions and equity on the planet."[77] Matthias Schmelzer, Andrea Vetter, and Aaron Vansintjan explain how degrowth addresses a combination of ecological, socio-economic, cultural, and feminist critiques as well as critiques of capitalism, of industrialism, and of South-North relations. In the process, it "has helped to politicize the debate on sustainability on development[...]."[78] Waste only enters these considerations as a function of production and consumption. The goal is to decouple production and consumption from human well-being, which automatically turns waste into a non-issue. This change in economic structure requires a psychological and cultural transformation of perspectives on key elements relating to modern life, redefining well-being in social rather than material terms (see Figure 7.3).

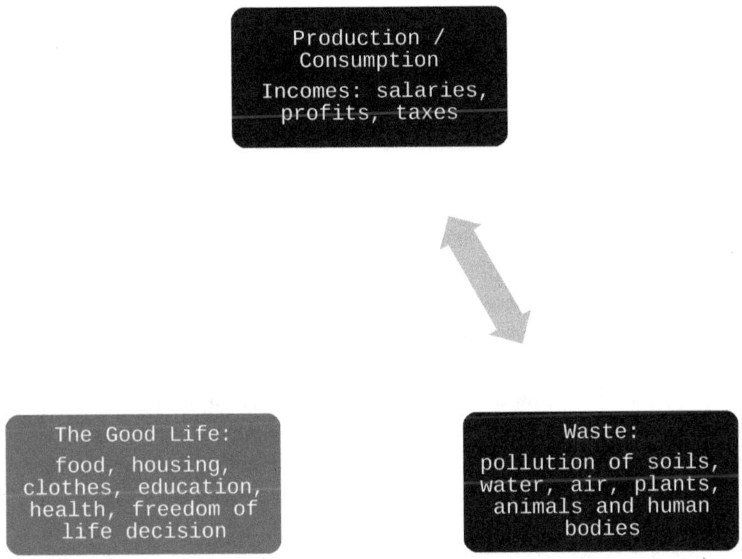

Figure 7.3 The degrowth decoupling of the good life from production and consumption.

Source: Copyright by Iris Borowy.

Given the substantially more ambitious scope, calls for specific transformational policies are both more diverse and more far-reaching than those for CE. The confinements of this paper do not allow a detailed comparison of plans that have been put forward by different authors. Nevertheless, core proposals that seem generally accepted within the degrowth community and are shared by key programmatic texts include the replacement of GDP with a different indicator of economic progress, a shift of development goals from labor to resource productivity and taxation from income to resources, the empowerment of local and self-governing structures, the transfer of funds from the North to the Global South to support ecological transitions in low-income countries, the reduction and regulation of trade, the transformation of productivity gains in work time reduction instead of profits, and legal caps for resource extraction, income, emissions, and waste generation.[79] Any implementation would obviously be a major transformation, not only of economies but of the way people organize their lives, interact with one another, and think about their environments and themselves.

The idea that perpetual economic growth is neither possible nor desirable has steadily gained ground within public debate.[80] In recent years, prestigious commissions, such as the Sarkosy-Stiglitz Commission in France or the Enquete Commission on Growth, Prosperity on Quality of Life of the German Parliament, have discussed problems of economic growth, and in July 2023 a large-scale meeting on degrowth, drawing 7,000 participants, took place in the European

Parliament.[81] However, its principle goes flatly against the central tenet of conventional economic doctrine and against many people's expectations, especially in times of crisis. Furthermore, it has not been formally endorsed by any government or organization, and informal implementations remain limited. This lack of implementation is partly related to the nature of the concept: its scope is vast, and many of the instruments listed above cannot be implemented on a small scale on a local or even a national level. For instance, a transformed tax system that would make labor-intensive products cheaper and resource-intensive products more expensive would provide environmentally useful incentives (a desired effect), but it would require governmental action and, realistically, also international tax harmonization since one country going alone would place its own exports at a disadvantage. Thus, implementing degrowth generates serious difficulties in an age of globalization and may necessitate a process of profound and extensive transformation on at least a regional level. One is tempted to say that the implementation of degrowth cannot happen slowly, which makes meaningful local experimentation and efforts at limited implementation difficult. Even within one country, the division of responsibilities might be difficult to coordinate, and some people, especially in democratic societies, might be put off by the paternalism entailed in a transformation that would involve many people radically changing the ways in which they live, work, and think. Critics of degrowth have drawn on this lack of easy appeal to conclude that it could only be implemented by a dictatorship and is, at best, "unrealistic utopianism."[82]

However, this weakness is also its strength, since degrowth provides a comprehensive concept with far-reaching solution potential for what most threatens wellbeing, enabling people around the world to enjoy good lives today and in the future. By emphasizing theoretical reconceptualization and national and international equity, it would place these transformations on a firm footing, and socio-economic changes would not be cosmetic but fundamental.

Degrowth, therefore, is much stronger in theory than CE but somewhat weaker in terms of practical experience, even in partial implementation. Local examples include transition towns[83] and local currencies.[84] In addition, a few projects have targeted "waste degrowth," sometimes with surprisingly conventional methods, such as the introduction of unit-pricing schemes in waste management in four Spanish communities.[85]

At present, the impact is ambivalent. Even though less directly concerned with waste than CE, degrowth does address the core point of rising quantities of waste: an economic system and a mindset aiming at increasing production and consumption as a way to increase well-being and fight off crises. In the long run, rejecting this line of thinking may be indispensable in fending off the ongoing waste crisis, but in the short run, the comparative radicalism of the concept is attractive to some people but scares away others and stands in the way of gradual implementation. The main argument of the degrowth movement is that a non-growing world is compatible with—or even necessary for—creating a world of widespread well-being, i.e., a world with the benefits currently made possible by the income generated through large-scale production, consumption, and waste. It sounds plausible but is empirically untested on a societal level. Existing evidence is ambivalent, but some suggests that rising incomes have diminishing returns on happiness and, depending on the circumstances, may even have negative effects.[86] However, there is also serious confirmation that rising incomes continue to increase well-being and that in societies with stagnant or declining consumption, happiness tends to decrease.[87] Happiness may not be the only criterion by which to evaluate development, or it may have to be weighed against the unhappiness caused by environmental destruction. There is a solid body of literature promoting proposals on how to create a degrowth system, and more plans can be expected from a recently awarded ERC grant of almost 10 million euros.[88] Nonetheless, plans are not as good as observable reality, and the absence of a tangible real-life example is a serious impediment to the quick spread of a concept that is out to disrupt and replace an existing system, the flaws of which are profound but known.

Interestingly, despite their theoretical contradictions, CE and degrowth are more compatible than may appear at first sight. CE focuses on making practical changes, with businesses as key partners and drivers of change, while degrowth begins by transforming ways of thinking about the economy and life in general, and business tends to be viewed as obstructionist. However, in their practical implementations in terms of waste, the concepts meet. Some considerations just seem too inherent to the problem at hand to escape similar conclusions. Thus, Patagonia management mused that "if materials cause pollution and energy causes pollution, but a business still needs to provide products to the customer, the obvious (albeit, less capitalist) solution to getting closer to circularity seems to be: Make less stuff."[89] Privileging repairs over buying new follows both CE and degrowth logic. The EU directive to establish a "right to repair" was proposed in March 2023 as part of an ostensibly intended transition to CE, but it could easily have come under a "degrowth label."[90] The most hopeful way forward may be found in countless grassroots initiatives such as local clothes swaps, flea markets, internet sale platforms like eBay, and platforms that offer discounts for buying new products when handing in old ones, such as *Yi Jiu Huan Xin* in China. On the business level, Globechain offers a marketplace platform for discarded items, advertising "waste a resource for everyone" while working on a commercial basis.[91] A growing list of influencers have individually begun living and promoting a life without waste, and they have a sizable following, both online and offline.[92] All these initiatives are helpful and fruitful, but they also share some degree of ambivalence regarding waste reduction policies, as they promote the practice while also engaging in commercial activities that will, at some point, inevitably result in waste. For all its commitment to CE, Patagonia still wants to (and needs to) sell pricey clothes and outdoor equipment. Even books on zero-waste management will eventually be waste.

Conclusion

Waste is the most visible manifestation of the contradiction that wealth based on mass production and consumption has both improved and harmed human lives. Different groups of people are clearly affected in different ways, with vulnerable groups experiencing more harm and less benefit. But generally speaking, people across the globe have been touched in some way by both the benefits and the harm of an economic system that systemically generates large quantities of waste. In that sense, waste provides the perfect evidence of the ambivalent nature of recent socio-economic development.

To date, no perfect strategy to break up this connection between the advantages and the disadvantages, the good life and hazardous waste, has been found. CE and degrowth aim at cutting the connection at different places, juxtaposing an unambitious, widely accepted concept with an ambitious notion that leaves many people suspicious. Some people may argue that it is better to have a mediocre policy that exists than a great policy that does not. Views on this question will depend on whether one believes that CE is able to achieve similar results (at least with regard to waste) to degrowth but without much of the ideological baggage, or whether one views the political component as the core of necessary transformations rather than baggage.

In theory, the combination of both concepts is difficult, since any acceptance of CE depends substantially on the promise that it will maintain or increase economic growth and the profits that go with it, which is precisely what degrowth seeks to prevent. On the other hand, there is a chance that practical projects will uncover so much common ground as to make these differences increasingly irrelevant. Both concepts aim at transforming waste material, with designs leading to waste generation as well as the underlying social and cultural structures and belief systems, so perhaps, eventually, it will not matter so much which of these transformations comes first. In this sense, addressing the waste crisis may not only transform materials and economic models, but also add new ambivalence to development models designed to transform industrialized societies. And this ambivalence may in turn open up fruitful pathways for the future.

Notes

1 "China unveils policy incentives to boost consumption," *CGTN*, July 31, 2023, https://news.cgtn.com/news/2023-07-31/China-unveils-policy-incentives-to-boost-consumption-1lT7JCK3ZlK/index.html.

2 Felix Richter, "China Is the World's Manufacturing Superpower," Statista, May 4, 2021, https://pdf.defence.pk/threads/china-is-the-worlds-manufacturing-superpower.733678/; "Lifting 800 Million People Out of Poverty – New Report Looks at Lessons from China's Experience," World Bank, April 1, 2022, www.worldbank.org/en/news/press-release/2022/04/01/lifting-800-million-people-out-of-poverty-new-report-looks-at-lessons-from-china-s-experience.

3 Flàvio F. Carmo, Andressa O. Lanchotti and Luciana H.Y. Kamino, "Mining Waste Challenges: Environmental Risks of Gigatons of Mud, Dust and Sediment in Megadiverse Regions in Brazil," *Sustainability* 12, no. 20 (2020): 8466, https://doi.org/10.3390/su12208466; "Mining waste," European Union, accessed July 29, 2023, https://environment.ec.europa.eu/topics/waste-and-recycling/mining-waste_en.

4 A quote attributed to Daniel Hoornweg, co-author of a programmatic World Bank report on waste, in Joseph Strombert, "When Will We Hit Peak Garbage?" *Smithsonian*,

October 30, 2013, www.smithsonianmag.com/science-nature/when-will-we-hit-peak-garbage-7074398/.

5 Jutta Bolt and Jan Luiten van Zanden, "The long view on economic growth: New estimates of GDP," in *How was Life: Vol. II, New Perspectives and Wellbeing and Global Inequality 1820*, ed. OECD (Paris: OECD 2021), 29-52, 44, www.oecd-ilibrary.org/sites/5fa1c6e1-en/index.html?itemId=/content/component/5fa1c6e1-en.

6 "Population by Income (%)," Gapminder, accessed July 23, 2023, www.gapminder.org/tools/#$model$markers$billy$encoding$selected$data$;;;;;;&chart-type=mountain&url=v1.

7 Muhammad Jami Husain, "Revisiting the Preston Curve: An Analysis of the Joint Evolution of Income and Life Expectancy in the 20th Century" (Keele University, 2011), www.researchgate.net/profile/Muhammad-Jami-Husain/publication/268408473_Revisiting_the_Preston_Curve_An_Analysis_of_the_Joint_Evolution_of_Income_and_Life_Expectancy_in_the_20_th_Century; Johan Mackenbach and Caspar Looman, "Life expectancy and national income in Europe, 1900–2008: an update of Preston's analysis," *International Journal of Epidemiology* 42, no. 4 (2013): 1100–1110, https://doi.org/10.1093/ije/dyt122.

8 NCD Risk Factor Collaboration, "A century of trends in adult human height," *eLife* 2016, 5:e13410, https://elifesciences.org/articles/13410.

9 Robert Fogel and Dora Costa, "A Theory of Technophysio Evolution, With Some Implications for Forecasting Population, Health Care Costs, and Pension Costs," *Demography* 34, no. 1 (1997): 49–66; Roderick Floud, Robert Fogel, Bernard Harris, and So Chul Hong, *The Changing Body: Health, Nutrition and Development in the Western World since 1700* (Cambridge: Cambridge University Press, 2011), 3–4.

10 John Helliwell, Richard Layard, Jeffrey Sachs, and Jan-Emmanuel De Neve, eds., *World Happiness Report* (New York: Sustainable Development Solutions Network 2020), 23.

11 See respective graphs in Our-world-in-data, https://ourworldindata.org/.

12 In addition to the citations above, for the GDP-waste nexus see: Manuel A. Zambrano-Monserrate, Maria A. Ruano, and Vanessa Ormeño-Candelario, "Determinants of municipal solid waste: a global analysis by countries' income level," *Environmental Science and Pollution Research* 28 (2021): 62421–62430, https://doi.org/10.1007/s11356-021-15167-9; David Meng-Chuen Chen, Benjamin Leon Bodirsky, Tobias Krueger, Abhijeet Mishra, and Alexander Popp, "The world's growing municipal solid waste: trends and impacts," *Environmental Research Letters* 15, no. 7 (2020), https://doi.org/10.1088/1748-9326/ab8659.

13 Jan Zalasiewicz et al., "Scale and diversity of the physical technosphere: A geological perspective," *The Anthropocene Review* 4, no. 1 (2017): 9–22, https://doi.org/10.1177/2053019616677743.

14 "International Day of Zero Waste," United Nations, accessed July 25, 2023, www.un.org/en/observances/zero-waste-day.

15 Silpa Kaza, Lisa C. Yao, Perinaz Bhada-Tata, and Frank Van Woerden, *What a Waste 2.0: A Global Snapshot of Solid Waste Management to 2050. Urban Development* (Washington, DC: World Bank, 2018), 1-3, 28.

16 Kaza et al., *What a Waste 2.0*, 17, 30.

17 Roland Geyer, Jenna R. Jambeck, and Kara Lavender Law, "Production, use, and fate of all plastics ever made," *Science Advances* 3, no. 7 (July 2017), https://doi.org/10.1126/sciadv.1700782.

18 C.P. Baldé, E. D'Angelo, V. Luda O. Deubzer, and R. Kuehr, *Global Transboundary E-waste Flows Monitor – 2022* (Bonn: United Nations Institute for Training and Research, 2022).

19 Kaza et al., *What a Waste 2.0*, 36–37.
20 C.P. Baldé et al., *The Global E-Waste Monitor 2017*, 11.
21 "International Day of Zero Waste," United Nations, accessed July 25, 2023, www.un.org/en/observances/zero-waste-day.
22 Kaza et al., *What a Waste 2.0*, 1–5, 18
23 D-Waste et al., *Waste Atlas: 2014 Report*, 4–13, www.atlas.d-waste.com.
24 Lucia Fazzo et al., "Hazardous waste and health impact: a systematic review of the scientific literature," *Environmental Health* 16, no. 107 (2017), https://doi.org/10.1186/s12940-017-0311-8.
25 Rob Nixon, *Slow Violence and the Environmentalism of the Poor* (Cambridge: Harvard University Press 2011).
26 "International Day of Zero Waste," United Nations, accessed July 25, 2023, www.un.org/en/observances/zero-waste-day (404).
27 Kaza et al., *What a Waste 2.0*, 1–5, 18.
28 Jennifer Clapp, *Toxic Exports: The Transfer of Hazardous Wastes from Rich to Poor Countries* (Ithaka and London: Cornell University Press, 2001); David Pellow, *Resisting Global Toxics* (Cambridge: MIT Press, 2007); Simone Müller, "Hidden Externalities: The Globalization of Hazardous Waste," *Business History Review* 93, no. 1 (2019): 51–74, https://doi.org/10.1017/S0007680519000357.
29 Kate O'Neill, *Waste* (Cambridge: Polity Press, 2019), chapter 4; Adam Minter, *Secondhand* (New York: Bloomsbury Publishing, 2019), chapter 11.
30 Adam Minter, "The Burning Truth Behind an E-Waste Dump in Africa," *Science*, January 13, 2016; "Guiyu, China – The Largest Electronic Waste Site in the World," WordAtlas, accessed July 22, 2023, www.worldatlas.com/articles/guiyu-china-the-largest-electronic-waste-site-in-the-world.html; Yvan Schulz and Anna Lora-Wainwright, "In the Name of Circularity: Environmental Improvement and Business Slowdown in a Chinese Recycling Hub," *Worldwide Waste* 2, no. 1 (June 2019): 1–13.
31 Patrick O'Hare, "Landfill life and the many lives of landfills," in *The Routledge Handbook of Waste Studies,* eds. Zsuzsa Gille and Josh Lepawsky (London and New York: Routledge, 2022), 55–67, quote 59.
32 Gary Haq and Alistair Paul, *Environmentalism since 1945* (London: Routledge, 2011), 7–25.
33 Kenneth E. Boulding, "The Economics of the Coming Spaceship Earth," in *Environmental Quality in a Growing Economy*, ed. Henry Jarrett (Baltimore: John Hopkins University Press, 1966), 3–14.
34 Mika Sillanpää and Chaker Ncibi, *The Circular Economy* (London: Academic Press, 2019), 4–7.
35 Used, for instance, here: www.recyclingconnections.org/post/five-steps-of-recycling;www.forbes.com/sites/stevebanker/2021/06/29/the-circular-supply-chain-a-push-for-sustainability/ or here: https://rankiapro.com/en/a-circular-economy-picking-the-easy-net-zero-fruit/.
36 "What is circular economy?" EPA, accessed July 24, 2023, www.epa.gov/recyclingstrategy/what-circular-economy.
37 Kaza et al., *What a Waste 2.0*, 120 off.
38 Kate O'Neill, *Waste* (Cambridge: Polity Press, 2019), 2.
39 "Circular Economy Introduction," Ellen MacArthur Foundation, April 4, 2023, https://ellenmacarthurfoundation.org/topics/circular-economy-introduction/overview.

40 "First circular economy action plan," European Commission, accessed July 21, 2023, https://environment.ec.europa.eu/topics/circular-economy/first-circular-economy-action-plan_en.

41 Olabode Emmanuel Ogunmakinde, "A Review of Circular Economy Development Models in China, Germany and Japan," *Recycling* 4, no. 27 (2019), https://doi.org/10.3390/recycling4030027.

42 "A circular economy for more sustainable resource use," Öko-Institut, accessed July 25, 2023, www.oeko.de/en/research-consultancy/issues/resources-and-recycling/translate-to-englisch-deutschland-europa-global-kreislaufwirtschaft-fuer-eine-nachhaltigere-rohstoffverwendung; Bundestag, "Reorganising the Law on Closed Cycle Management and Waste," February 24, 2012, www.bmuv.de/fileadmin/Daten_BMU/Download_PDF/Abfallwirtschaft/kreislaufwirtschaftsgesetz_en_bf.pdf.

43 BMUV, "Eckpunkte der Novellierung des Kreislaufwirtschaftsgesetzes (KrWG)," October 28, 2020, www.bmuv.de/themen/wasser-ressourcen-abfall/kreislaufwirtschaft/abfallpolitik/uebersicht-kreislaufwirtschaftsgesetz/eckpunkte-der-novellierung-des-kreislaufwirtschaftsgesetzes-krwg.

44 "Abfallbilanz 2020," Destatis, accessed July 28, 2023, www.destatis.de/DE/Themen/Gesellschaft-Umwelt/Umwelt/Abfallwirtschaft/Tabellen/liste-abfallbilanz-kurzuebersicht.html#613014.

45 Jianguo Qi et al., *Development of Circular Economy in China* (Singapore: Springer, 2016), xxvi.

46 Keyou (Emma) Feng and Chun-Yin (Anson) Lam, "An Overview of Circular Economy in China: How the Current Challenges Shape the Plans for the Future," *The Chinese Economy* 54, no. 5 (2021): 355–371, https://doi.org/10.1080/10971475.2021.1875156.

47 Jouni Korhonen, Antero Honkasalo and Jyri Seppälä, "Circular Economy: The Concept and its Limitations," *Ecological Economics* 143 (2018), 737–746.

48 "Mexico City's Circular Economy Law Approved," Global Factor, February 23, 2023, www.globalfactor.com/en/mexico-citys-circular-economy-law-approved/.

49 Kaza et al., *What a Waste 2.0*, 120.

50 Regional 3R and Circular Economy Forum in Asia and the Pacific, https://uncrd.un.org/content/regional-3r-and-circular-economy-forum-asia-and-pacific; World Circular Economy Forum 2023, https://wcef2023.com/about/.

51 OECD, "Towards a more resource-efficient and circular economy: The role of the G20," 2021, www.oecd.org/environment/waste/OECD-G20-Towards-a-more-Resource-Efficient-and-Circular-Economy.pdf; World Bank, "Transitioning to a Circular Economy," March 17, 2022, https://ieg.worldbankgroup.org/evaluations/transitioning-circular-economy.

52 See country data in https://ourworldindata.org/grapher/total-waste-generation?, accessed July 31, 2023.

53 The World Bank, *Squaring the Circle: Policies from Europe's Circular Economy Transition* (Washington: World Bank, 2022), 4.

54 Adam Minter, *Junkyard Planet* (New York: Bloomsbury Press, 2013), 253–254.

55 Feng and Lam, "An Overview of Circular Economy in China."

56 Yvan Schulz and Anna Lora-Wainwright, "In the Name of Circularity: Environmental Improvement and Business Slowdown in a Chinese Recycling Hub," *Worldwide Waste* 2, no. 1 (June 2019): 1–13.

57 O'Neill, Waste, 7.

58 Baldé et al., *Global Transboundary E-waste Flows Monitor – 2022*, 15.

59 Mark Weick and Nicole Ray, "How companies can leverage the circular economy to address global e-waste," Ernst & Young, February 20, 2023, www.ey.com/en_us/climate-change-sustainability-services/how-circular-economy-models-can-address-global-e-waste.

60 Randeep Somel, "A Circular Economy: Picking the Easy Net-Zero Fruit," July 27, 2021, https://rl360adviser.com/news/2021/fund-article-mandg-a-circular-economy-sep21.htm.

61 Federico Savini, "The economy that runs on waste: accumulation in the circular city," *Journal of Environmental Policy & Planning* 21, no. 6 (2019): 675–691, https://doi.org/10.1080/1523908X.2019.1670048.

62 O'Neill, Waste, 17; Minter, Junkyard Planet, 259.

63 Adam Minter, *Junkyard Planet* (New York: Bloomsbury Press, 2013), 254, cursive in the original.

64 Menmeng Cui, "Key Concepts and Terminology," in *An Introduction to Circular Economy,* eds. Lerwen Liu and Seeram Ramakrishna (Singapore: Springer, 2021), 26–30.

65 Weick and Ray, "How companies can leverage the circular economy to address global e-waste."

66 Cui, "Key Concepts and Terminology," 26–30.

67 Kersty Hobson and Nicholas Lynch, "Diversifying and degrowing the circular economy: radical social transformation in a resource-scarce world," *Futures* 82 (2016): 15–25.

68 "Collect, recirculate and recycle," H&M, accessed July 26, 2023, https://hmgroup.com/sustainability/circularity-and-climate/recycling/; "Made to be Remade," Adidas, accessed July 30, 2023, www.adidas.com/us/sustainability/wear-longer.

69 Dalia D'Amato, Jouni Korhonen, and Anne Toppinen, "Circular, Green, and Bio Economy: How Do Companies in Land-Use Intensive Sectors Align with Sustainability Concepts?" *Ecological Economics* 158 (2019): 116–133.

70 Archana Ram, "Our Quest for Circularity," March 10, 2021, www.patagonia.com/stories/our-quest-for-circularity/story-96496.html.

71 John Ehrenfeld and Mariam Chertow, "Industrial symbiosis: the legacy of Kalundborg," in *A Handbook of Industrial Ecology,* eds. Robert Ayres and Leslie Ayres (Cheltenham: Edward Elgar, 2002), 334–350.

72 Savini, "The economy that runs on waste," 687.

73 Ibid., 688.

74 Dennis Meadows et al., *Limits to Growth* (New York: Universe Books, 1972).

75 Nicholas Georgescu-Roegen, *The Entropy Law and the Economic Process* (Cambridge: Harvard University Press, 1971); Herman Daly, *Steady-State Economics: The Economics of Biophysical Equilibrium and Moral Growth* (New York: W. H. Freeman, 1977).

76 Barbara Muraca and Matthias Schmelzer, "Sustainable degrowth: Historical roots of the search for alternatives to growth in three regions," in *History of the Future of Economic Growth: Historical Roots of Current Debates on Sustainable Degrowth*, eds. Iris Borowy and Matthias Schmelzer (London and New York: Routledge 2017), 174–197.

77 "Definition," Research and Degrowth, April 8, 2023, https://degrowth.org/definition/.

78 Matthias Schmelzer, Andrea Vetter and Aaron Vansintjan, *The Future is Degrowth* (London/New York: Verso, 2022), 1.

79 Inês Cosme, Rui Santos and Daniel W. O'Neill, "Assessing the degrowth discourse: A review and analysis of academic degrowth policy proposals," *Journal of Cleaner Production* 149 (2017): 321–334; Herman Daly, "Five policy recommendations

for a sustainable economy," *Feasta Review*, accessed July 20, 2023, http://feasta.org/documents/feastareview/daly2.htm; Tim Jackson, *Prosperity without Growth* (London: earthscan, 2011), 173–185; New Economics Foundation, *The Great Transition* (London: NEF, 2010); Schmelzer, Vetter and Vansintjan, *Future*.

80 For a brief list of references to degrowth in newspapers, see https://degrowth.org/degrowth-in-the-news/, accessed July 21, 2023.

81 Joseph E. Stiglitz, Amartya Sen and Jean-Paul Fitoussi, "Report by the Commission on the Measurement of Economic Performance and Social Progress," https://ec.europa.eu/eurostat/documents/8131721/8131772/Stiglitz-Sen-Fitoussi-Commission-report.pdf; Frederike Strunk, "The German Enquete Commission on growth, prosperity and quality of life: a model for futures studies?" *European Journal of Futures Research* 3 (2015), https://doi.org/10.1007/s40309-014-0058-1; Myriam Best, "Beyond Growth 2023: critical reflections on a 'historic conference' with blind spots," *Resilience*, July 26, 2023, www.resilience.org/stories/2023-07-26/beyond-growth-2023-critical-reflections-on-a-historic-conference-with-blind-spots/.

82 Martin Wolf, *The Crisis of Democratic Capitalism* (New York: Penguin, 2023), 223.

83 "A movement of communities coming together to reimagine and rebuild our world," Transition Network, accessed July 21, 2023, https://transitionnetwork.org/.

84 Kristofer Dittmer, "Local currencies for purposive degrowth? A quality check of some proposals for changing money-as-usual," *Journal of Cleaner Production* 54 (2013): 3–13.

85 Gabriel Weber, Ignazio Cabras, Maria Calaf-Forn, Ignasi Puig-Ventosa, Giacomo D'Alisa, "Promoting Waste Degrowth and Environmental Justice at a Local Level: The Case of Unit-Pricing Schemes in Spain," *Ecological Economics* 156 (2019): 306–317.

86 John Helliwell, Richard Layard and Jeffrey Sachs, eds., *World Happiness Report* (York: UN Sustainable Development Solutions Network, 2012), 5, https://s3.amazonaws.com/happiness-report/2012/World_Happiness_Report_2012.pdf.

87 Fanning Devers and Daniel O'Neill, "The Wellbeing–Consumption paradox: Happiness, health, income, and carbon emissions in growing versus non-growing economies," *Journal of Cleaner Production* 212 (2019): 810–821.

88 Nathan Barlow, "Degrowth awarded an ERC Grant: An interview with Giorgos Kallis," *Resilience*, December 1, 2022, www.resilience.org/stories/2022-12-01/degrowth-awarded-an-erc-grant-an-interview-with-giorgos-kallis/.

89 Ram, "Our Quest for Circularity."

90 "Proposal for a Directive of the European Parliament and of the Council on common rules promoting the repair of goods and amending Regulation," March 22, 2023, https://commission.europa.eu/system/files/2023-03/COM_2023_155_1_EN_ACT_part1_v6.pdf.

91 Globechain, accessed July 29, 2023, https://globechain.com/.

92 "The Top Ten Zero-Waste Influencers in the World Today," Zerowaste, April 26, 2021, www.zerowaste.com/blog/the-top-ten-zero-waste-influencers-in-the-world-today/.

8 Farming the Wind

Aeolian Politics and the Sacred Desertscapes (Orans) of Rajasthan

Dibyadyuti Roy and Nisha Paliwal

Introduction

In 2021, the *International Energy Agency*—the global autonomous intergovernmental organization that provides policy recommendations, analysis, and data on the entire global energy sector—published the "*India Energy Outlook Report*" that noted

> tremendous opportunity for India to develop and successfully meet the aspirations of its citizens without following the *high-carbon pathway* that other economies have pursued in the past. Seizing this opportunity is critically important for India, and critically important for the world.[1]

The rhetoric employed in this report highlights the ambivalence that has been central to India's energy aspirations since its independence from British rule in 1947. On one hand, it acknowledges India's projected status as the country with the largest increase in energy demand by 2040, signaling its formidable energy future. At the same time, the report is a coercive gesture that assumes India's alleged duty toward the global order: of embracing sustainable energy systems and eschewing carbon-intensive development. This duality, which pits India's burgeoning energy needs against its expected role of embracing sustainable energy solutions, epitomizes the conflicting forces that have shaped postcolonial India's energy ambitions. Being an erstwhile colonized state that became the unwilling laboratory for many extractive and techno-positivist British imperial experiments, there remains skepticism in postcolonial India's imaginary about uncharted scientific and technological advancement.[2] Consequently, independent India's energy needs have often been caught up in the dilemma between avoiding Western models of consumerist capitalism while also needing to meet nationalist aspirations toward an energy-driven postcolonial modernity. As a nation that occupies second position (after China) with regard to the number of Clean Development Mechanism (CDM)[3] projects, India's commitment to sustainable development through renewable energy infrastructures is problematized by the

DOI: 10.4324/9781032627984-11

lack of a "detailed understanding of the governance of CDM projects at the local level, and their long-term social, economic and environmental impacts on the communities drawn into the CDM's global reach."[4] Particularly the devolution of accountability at the local level in relation to the lived experiences of both human and non-human communities, which has often been a troubling feature of global CDM projects, is exemplified by the ambivalent *aeolian politics*[5] at play within the Thar Desert Ecosystem in India. In this chapter, we extend and extrapolate Boyer and Howe's recent conceptualization of this term (aeolian politics), which is applicable to contexts where "renewable energy can be installed in ways that do little to challenge the extractive logics that have undergirded the mining and fossil fuel industries."[6] By doing so, we hope to understand how an uncritical assessment of renewable energy infrastructures can elide the complex nexus of political motivations, neoliberal ideologies, and private interests that very often govern such projects, especially in the Global South(s)/Majority Worlds.[7] In addressing the key questions raised by this volume around the ambivalences of ecological transformation, this chapter also examines the tangible and intangible fallout from the proliferation of wind farms in Jaisalmer and the adjoining desert ecosystems in Rajasthan, India, and specifically focuses on a particular case study, namely *Orans* that are community-conserved areas in Rajasthan and considered sacred by local inhabitants. These sites

> are rich in biodiversity[...] where herders take their livestock for grazing [and] are places for communal congregations, festivals, and other social events, the performance of which is linked to agrarian rhythms and the continued commitment of the communities towards environmental conservation.[8]

Through interrogating how the physical and discursive positioning of *Orans*—as community-oriented spaces of indigenous knowledge—is being threatened through aeolian politics, this chapter illustrates the coercive potential of renewable energy imaginaries and infrastructures in altering "a *place*[...] identified as relational, historical and concerned with identity" into a *non-place* "which cannot be defined as relational, or historical, or concerned with identity."[9]

Methodologically, our intervention combines a brief survey of the exigencies of renewable energy in postcolonial India, leading to a contextual close reading of contemporary policy discourses around wind power in the country, with ethnographic engagements—framed by oral testimonials from stakeholders, affected by wind farm infrastructures around *Aalaji ka Oran* in Jaisalmer, Rajasthan. By emphasizing the centrality of renewable energy, its legacies, politics, and future within narratives of sustainable development in postcolonial India, and the resultant impact on minority communities and indigenous knowledge systems, this chapter situates aeolian politics in the nation as a vital site for locating the ambivalences of ecological transformation, which are coextensive with similar corporate wind economies around the world.

The Politics of (Renewable) Energy in Postcolonial India: Ambivalences, Ambiguities, and Tensions

In this section, we look at the ambivalences, ambiguities, and tensions characterizing the socio-technical imaginaries of India's renewable energy infrastructures, as well as the phenomenon of green grabbing[10] that performs the material and discursive erasure of the concerns of local populations about their sacred sites, which are endangered by the uncritical expansion of renewable energy infrastructures. After India's independence from British rule in 1947 and following the traditional model of development in postcolonial nations of/in the "Global Majority,"[11] there was a concerted effort toward increased agricultural production and rapid industrialization, both of which were seen as bulwarks of a developing economy. Such models of development hinged on two key factors, namely, self-sufficiency in energy production as well as providing energy security infrastructures that would contribute to the political independence of a newly liberated nation. Accordingly, the newly constituted Planning Commission in post-independence India put the country's energy needs at the core of its developmentalist ambitions.[12] However, the immediate situation in a newly independent country that had been economically, socially, and psychically ravaged by two hundred years of colonial rule, as well as the violent aftermath of a bloody partition, meant that developmental ambitions of the First Five-Year Plan (1951–1956) "*could not but be a modest one*, even if the planners and leaders[…] wanted a more ambitious approach (our emphasis)."[13] Therefore, even though energy considerations did not assume central importance within the first Five-Year Plan, there was considerable debate amongst Indian national leaders around the wasteful use of cattle dung as fuel and the rapid rate of deforestation.[14] Such conversations led to the Ministry of Food and Agriculture's resolution in the National Forest Policy (1952) to reduce the use of cattle dung as household fuel, with a country-wide study in 1953 concluding the need for soft coke[15] as a possible alternative. However, the India of this period was constrained by both a lack of investment required to support its logistical needs (distribution) and the reluctance of a largely rural population to move toward a priced commodity (such as soft coke) as opposed to firewood and cow dung, which were available for free or at minimal private cost. In the following years, the Energy Survey Committee (ESC) (1962), which emerged from the policy conversations of the Second Five-Year Plan (1956–1961), was one of the first consolidated attempts to map the energy policy and resources available for shaping India's developmental aspirations.[16] However, both the ESC and its report had a minimal impact due to two primary limitations: first, "most of its recommendations were hardly considered with any seriousness,"[17] even by the consequent Five-Year Planning Commissions, since the ESC was composed largely of foreign advisors who were supported by a small Indian staff contingent. Second, most of the energy policy and supply recommendations made by the ESC were predicated on predetermined ideologies of developmentalism rather than granular energy considerations, which should ideally determine developmental policies.

A paradigm shift in terms of energy policies took place in India following the 1973 global oil crisis, which saw the nation's oil bills move from 20% to 80% of its export earnings.[18] This exigence led to the reframing of India's energy policies predicated on the difficult choices between "a disastrous, drastic slowdown of growth or of maintaining the same level of growth (if not a higher level) at any cost."[19] Since the developmental aspirations of a postcolonial nation playing catch-up with developed Western economies could not be curtailed, India chose the latter path. Importantly, while the reliance on dominant coal-based energy infrastructures increased in the aftermath of the oil crisis, there was also a movement toward exploring renewable energy options, which led to the establishment of the Commission for Additional Sources of Energy (CASE) in 1981. This commission was the precursor to the Department of Non-Conventional Energy Sources (DNES), formed in 1982 and subsequently replaced by the Ministry of Non-Conventional Energy Sources (MNES) in 1992.[20] The activities of CASE and the establishment of MNES in 1992 were influenced by increasing global discussions on environmental protection that began in the 1970s. These discussions also prompted the Indian Supreme Court to recognize the environment as "a part of the Fundamental Rights of every [Indian] citizen."[21] Over the last three decades, India's commitment to renewable energy (solar, hydro, wind, and bioenergy) has been shaped by rising electricity consumption, rapid industrialization, and its vulnerability to climate change. These factors have made the country an opportune site for exploring wind-power, and it is currently "the fourth largest producer of wind power in the world [...] with a strong manufacturing base."[22] However, as is the case with most countries in the Global South/Majority Worlds, the politics of renewable energy and the proposed energy transition(s) are fraught with "little or no social, political or economic transition attached to it."[23] The following sections discuss the infrastructural and discursive challenges central to aeolian politics in Rajasthan, India. Using the case study of *Orans*, this intervention illustrates how "dreams of Aeolian futures are [often] paradoxical; what are heavenly images for some are nightmares for others."[24]

Revitalizing Orans: Challenging Coloniality through Interrogating Wasteland Narratives

> Once upon a time, there was an ocean here. Its waves would come pursuing each other. One does not know why and how the waves of time dried up the incommensurable ocean. Now there is an ocean of sand. Even today, waves continue to pursue each other, except now they are waves of sand instead of water.[25]

Mythically, the river Sarasvati flowed through Rajasthan during the Indus-Sarasvati civilization. However, it is believed that subsequent tectonic shifts led to the disappearance of the river underground and led to the gradual desertification of the land, which later became the Thar Desert, stretching from the Aravallis in India to the Indus in Pakistan.[26] Jaisalmer, the largest district of Rajasthan with an area of 38,401 sq. km, receives an average annual rainfall of about 160 mm, thereby making it difficult to harvest crops as a primary source of livelihood. Therefore, cattle rearing has become the main occupation of most communities in this region, along with a traditional method of agriculture, called *khadeen*[27] farming in some areas, which is entirely based on the principles of diverting and collecting run-off water into low-lying land through a catchment. As the rainwater collects and seeps into the soil, thus enabling moisture conservation, specific crops such as gram and mustard that require less water are grown without further irrigation practices. Contextually, *Orans*, also known as *devbanis* (God's Forest) in some areas, are community-conserved areas (CCAs) that form important sources of water as well as grazing ground in the desert. Dominant developmentalist logic, which frames deserts as wastelands, fails to acknowledge the ontological understanding of the word *Oran*, which originates from the Sanskrit term *aaranya*, meaning "forest," and firmly situates *Orans* within a broader indigenous conception of Nature. This understanding is not constrained by the limiting hegemonic imaginaries of normative environmentalism. The chapter highlights how the developmentalist perspectives accompanying wind energy infrastructures in India overlook this deeper cultural and spiritual significance of *Orans* for local communities, as they are rooted in a holistic and indigenous understanding of nature, which transcends the often narrow conceptions of Western environmentalism. Significantly, while *Orans*, as vast tracts of desert land, have historically symbolized a referential "place" that is "relational, deeply historical and intimately connected to identity, both social and individual,"[28] their categorization as wastelands by the National Commission on Agriculture (NCA) has discursively altered them into "non-places," depriving these sites of either any identity or an intimate relationship with the local community. As repositories of bio-genetic diversity, *Orans* also act as ecological warehouses for protecting endemic, endangered, or vulnerable species, medicinal plants, and so forth. Owing to a complementary relationship that exists between cattle and biodiversity, the composition of *Orans* has evolved in response to the grazing practices and requirements of livestock.

Significantly, these sites are revered through rituals that honor the deities believed to reside there, and these rituals are accompanied by strict rules governing resource use. At the heart of this traditional biodiversity management system is a local folk deity protecting their domain and who is honored by a ritual that involves the process of circumambulation and pouring of milk around the territorial boundary, traditionally known as "*Dudh ki dhar dena*." This ceremony establishes an ecological understanding and symbiotic relationship by acknowledging the vital connection between water, vegetation, and the livestock-driven economy—through sanctifying the territory as sacred. However, the ecological balance and biodiversity and/in *Orans* are currently threatened by India's energy demands in terms of finding space for large-scale projects that also subscribe to the 2030 United Nations Sustainable Development Goals for meeting 50% of global energy demands through renewable energy (Renewable Energy in India—Indian Power Industry Investment, n.d.). The state of Rajasthan and its *Orans* has emerged as a key participant due to the availability of vast tracts of arid lands.[29]

According to a study conducted by the Indian Institute of Technology, Jodhpur, Rajasthan, the long-perceived "barren wasteland" of Thar Desert harbors "remarkable biodiversity" with "682 species of flora and 1,195 species of fauna" but is often neglected.[30] In a report published by the Institute of Energy Economics and Financial Analysis, the state of Rajasthan emerges as a significant location, with a desert of approximately 102,000 sq. km and an abundance of "barren land."[31] While the categorization of "wastelands" in revenue records in India has existed since the precolonial era, this aggressive process of categorizing local community desert spaces as "wastelands" arguably emerged from the global shift toward ecological modernization during the 1970s. Encroachment in such spaces is primarily governed by a determinist environmental paradigm that carries forward an imperial legacy of justifying colonial models thereof that promise "improvement" and "civilization."[32] Building on the above premise, this section highlights how large-scale renewable energy projects in Jaisalmer, Rajasthan, enforce green grabbing[33] by enclosing these vast tracts of community-conserved areas, known as "*Orans*." This enclosure and the disruption of the pastoral economy illustrate the infrastructural colonization performed by green energy projects in the Thar Desert, which are legitimized by hegemonic imaginaries of wastelands.

The term "wasteland" finds its genesis in the text "Two Treatises of Government" by John Locke, in which he stresses the need for intensive cultivation and agricultural productivity upon "individually enclosed" lands that transcend "a state of nature" to "nationhood" and "a state of civilization."[34] The imperative to mitigate the climate crisis through renewable energy forms is an acute concern in the Global South, where green grabbing of commons is disrupting the ecological balance and, in this case, threatening long-preserved *Orans*. The parallels of Locke's conception of wasteland as opposed to fertile and value-producing land influence a binary opposition of "waste" and "green" in contemporary models of renewable energy implementation, which in turn governs the categorization of landscapes with the notion of the "higher and best possible use of land." Large stretches of "agrarian and common lands have been enclosed or degraded to facilitate the global proliferation of large-scale renewable energy infrastructure."[35] The perception of the Thar Desert as "a curse and unfit for human habitation" stems from the influence of corporate and state-sponsored institutions that objectify and categorize such spaces through the lens of supermodernity.[36] This view has turned the Thar Desert from a physical referential place into a conceptual "non-place" in policymaking circles.

Such bureaucratic imaginaries can be historically located in India's post-independence developmentalist agenda, which valued regulation and state control over sustained community engagement. In the decades following India's independence, the pursuit of modernization and progress was heavily reliant on large-scale projects, such as the construction of dams and the centralized control of river valleys for irrigation purposes. These ambitious undertakings, however, frequently disrupted the harmonious relationship between local communities and their natural environments. Furthermore, they interfered with the interdependence of common resources and traditional management systems[37] that had been established over time, thereby eroding the delicate balance that had existed between the people and their surroundings.[38] The initiation of such hegemonic imaginaries for wasteland classification in postcolonial India can be traced to the establishment of the Nation Wasteland Development Board (NWDB) in 1985, which aimed to address the ecological and socio-economic challenges arising from poverty and deforestation. The NWDB's primary objective was to mitigate these issues by promoting the cultivation of fodder and fuelwood on lands that were previously considered unproductive or wastelands,[39] and it therefore focused on classifying lands based on their productive potential, with the aim of identifying sites that were considered suitable for productive utilization or "improvement" according to their hegemonic criteria. Carrying forward these longstanding approaches to contemporary times, the National Remote Sensing Centre (NRSC) recently, in 2019, produced the Wasteland Atlas of India, which estimates "the spatial extent of wastelands across the country."[40] The primary objective of such efforts is to abstract and categorize wastelands through the dominant lens of productivity, classifying them into 32 categories and presenting this information through a range of tables and maps. Complementing this initiative, the Energy Department of the Government of India introduced The Rajasthan Wind and Hybrid Energy Policy in 2019. This policy recognizes the "vast and barren landscapes in western Rajasthan" and aims to provide "a new impetus to the already vibrant wind power sector in the State." It also presents an opportunity for existing and new investors "to exploit the renewable potential for which Rajasthan is exceptionally well positioned."[41] According to Objective 5.4 of the document, the policy aims at the "productive use of abundant wasteland, thereby utilizing the unutilised/under-utilised land for the creation of wind energy hubs."[42] Additionally, according to the data presented in the Wasteland Atlas over the past few years, the total wasteland area (TWA) in Rajasthan has declined due to various factors, including the increasing number of green energy projects. In 2003, the TWA of Jaisalmer was 38,401 sq. km, which was 76% of the Total Geographical Area (TGA). However, this figure rapidly declined in successive years, with 26,244 sq. km in 2005 and 25,059 sq. km in 2008. This is mediated through ideological agendas of development that erase previous property regimes and clear land for dominant imaginaries of sustainable development. Thus, the construction of *Orans* as a "non-place" for the interests of energy development follows the "discursive logics of (neo)colonial and green energy land politics,"[43] which reveals "how much energy and climate issues are framed in [...] technical, simplified and depoliticized ways that deliberately elude questions of dispossession, ownership, property rights or inclusive development."[44] Unfortunately, these categorizations draw heavily from a colonial legacy that defines development through an "(un)making of space for capital accumulation"[45] alongside a decided indifference to its impact on local communities and ecological biodiversity. As Reed points out, this legacy of developmentalism is inherently linked to an "aestheticizing, pastoral form of environmentalism," whereby lush, green valleys are celebrated as representations of Nature, while desert regions are rarely recognized within such dominant imaginaries.[46]

Understanding the movement of policy around wastelands, its rhetoric, and the implications for narrativizing desert areas as wastelands provides a clear understanding of statist objectives: understanding energy transition as a "salvational object"[47] at "places that are relatively remote from the centres of consumption."[48] This dynamic indeed becomes the core of "aeolian politics" in Jaisalmer.

Contrary to the portrayal of commons as "wasteland" or unproductive areas, the following section, which emerges from a broader study grounded within various *Orans* in the Jaisalmer district of Rajasthan, will illustrate how the community landscape of *Aalaji ka Oran*, is perceived and experienced by local village inhabitants, thereby presenting counter-imaginaries that challenge dominant corporate and statist notions. These alternative perspectives shed light on the multifaceted relationships and intrinsic value that communities ascribe to their surrounding environments, which extend beyond the narrow lens of productivity and economic utility. Such narratives highlight how "wastelands" have played a crucial role throughout generations by supporting livelihoods through offering regenerative vegetation, meeting the fodder needs of livestock, and forming a symbiotic relationship that justifies why conserving *Orans* must remain a key ecological adaptive strategy in the harsh desert environment of the Thar ecosystem.[49]

Reclaiming Aalaji ka *Oran* in Jaisalmer: Recognizing Local Tensions Amidst Global Concerns

The Thar Desert of Rajasthan consists of nearly 5,370 sq. km of *Orans*, which play a significant role in supporting livelihoods by promoting a livestock-based economy in the desert.[50] However, they are threatened by the escalating number of hybrid (solar and wind) energy projects on these common lands. The global land rush for large-scale renewable energy is an acute concern when common lands take the brunt of climate change mitigation efforts, which disregard the dynamic socio-cultural practices of places, especially in the Global South. As per the law promulgated by the Ministry of Environment and Forests in India, wind energy projects in the country are exempted from environmental impact assessments (EIAs), which further widens the divergence between the imaginaries of local inhabitants and "eco-capitalists."[51] For indigenous communities, "the proliferation of machines from sprinkle to saturation is not an engineering project. It is a social, political and aesthetic struggle"[52] in which expansive tracts of common lands are currently being "enclosed or degraded to facilitate the global proliferation of large-scale renewable energy infrastructures."[53] In a bid to meet India's sustainable development goals, which include achieving a target of 500 GW of non-fossil, fuel-based energy by 2030, as set at COP26, the arid lands in Jaisalmer are experiencing the effects of infrastructural colonization. Despite lacking "visible or physical boundaries, walls, or limits,"[54] the proliferation of windmills in Jaisalmer extends physically through the construction of roads, electric towers, transmission lines, storage stations, and more. In this section, we specifically delve into *Aalaji ka Oran*, situated approximately 25 kilometers from the city of Jaisalmer—a region grappling with the expansion of the Wind Energy Project in Tejuva, Rajasthan.[55] As discussed in the previous section, every *Oran* has its own set of established rules and regulations in place to protect ecology and biodiversity, including endangered plant species, medicinal plants that are used to treat livestock, and plant species such as *Khejari* (*Prosopis cineraria*), *Babul* (*Acacia nilotica*), and *Ber* (*Zizyphus mauritiana*) used for grazing animals. These established customs may range from completely restricting the collection of any resources from the desertscapes to a localized ban on felling specific species of trees that have sacred significance to the community. Acting as an intermediary between the local people and the folk deity is a *bhopa*, who resides in the shire of the folk deity and takes care of the *Oran*. The *bhopa* is the central figure in a shire and serves as an oracle, connecting the local community to their deity and serving as a conduit of communication by mediating on the everyday problems of local inhabitants. Although it is challenging to demarcate a precise boundary of *Aalaji ka Oran*, the local community possesses an intimate familiarity with its territory despite the absence of any physical demarcation, and this profound connection is deeply ingrained in their cultural practices. The temple of Aalaji in Salkha village serves as a religious symbol, signifying the land as consecrated and a sacred marker of the *Oran*. According to legend, Aalaji was a warrior who lived in the Kanoi village of Jaisalmer and was the second son of Bhansi Ji. During the year 1550, when Maharawal Lunkaran ruled Jaisalmer, Prince Pathan Amir Ali of Kandahar took refuge and camped in Kishanghat, two kilometers away from the fort of Jaisalmer. Amir Ali realized that there was a lack of a permanent army and therefore informed Maharawal that his wives desired to meet the queens at the seraglio. Deceiving the king, he entered the fort with soldiers hiding in chariots and took over the throne. Intuitive about the crisis that had befallen, Aalaji tightened the saddle on his horse and left for Jaisalmer, and upon reaching it, he saw the siege of the Mughal army around the fort. While fighting the army, Aalaji was badly injured, and as he was just about to be beheaded, he called upon his Guru (spiritual guide), and a pair of eyes supposedly opened in his chest, which allowed him to keep fighting for seven days and seven nights. All this while moving from Jaisalmer to Derawar (now in Pakistan), until the earth burst open and Aalaji was absorbed into it. Though there are several temples dedicated to the folk deity Aalaji, the primary temple in India associated with him is in the Salkha village and is called "Chhatrapati Shri Aalaji Sheshan Mandir." This location marks the territory where Aalaji's decapitated head supposedly came to rest. Another significant temple, situated in Pakistan, commemorates the place where the earth engulfed his body. The idol in this temple depicts Aalaji as a headless warrior mounted on his horse. With the predominance of oral customary practices, the identification of written documentation related to land tenure proves challenging. The region around Aalaji's temple has been designated as an "*Oran*" by the local community. This area is consecrated to honor the deity, with the long-standing tradition rooted in ecological sustainability. The status of an "*Oran*" protects the land, safeguards biodiversity, and sustains the pastoral livelihood of the community. Due to the deep-seated reverence for *Orans*, community members refrain from felling trees or even plucking a leaf from the vegetation in this area. During an interview undertaken for the broader project with Chattar Singh Jam[56] from Ramgarh village, this sentiment was eloquently expressed when he noted, "*Oran sirf zameen nahi hai...Prakriti bolti hai..uski awaz hai* (The *Oran* is not just land. Through it the voice of Nature can be heard)"[57]

During the initial phase of the Wind Energy Project in Tejuva in 2012, considerable opposition arose within the local community, and the project was consequently portrayed as a means to supply electricity to the neighboring villages. Despite objections, the construction of the project began with the assurance that the windmills would be temporary infrastructure that would be removed within a few years. However, local residents expressed their preference to forgo electricity if it meant preserving the land that supported their entire livelihood. The infrastructural colonization unfolded gradually, and as one of the respondents interviewed, Reyan Khan[58] notes, "Though there isn't a wired boundary prohibiting entry into the area, the windmills are completely surrounding us and have taken over our land (sic)." He further adds that even the cattle are reluctant to approach the windmills due to the unsettling loud sounds they produce. To tackle this encroachment on the *Orans* of Jaisalmer, a collective known as "Team *Oran*" has emerged that comprises the local villagers and is led by community leader and activist Sumer Singh Bhati of the Sanwta village in Jaisalmer. This group has taken on the responsibility of raising awareness about *Orans* among the younger generation and combating the encroachment by green energy infrastructures, which poses a threat to local biodiversity. The high-tension lines and windmills associated with this infrastructure result in the daily demise of numerous birds. Team *Oran* actively organizes protests (see Figure 8.1) and movements to preserve and protect sacred *Oran* lands. Their efforts include conducting *parikrama* (circumambulation) around the *Orans* as an act of reverence to the deities. They also submit memorandums and demands to the

Figure 8.1 Villagers gathered for Oran Yatra in Degrai Oran on October 17, 2023.

Source: Copyright by Sachin.

Chief Minister, calling for the proper categorization of *Oran* in government land records. Team *Oran*'s ongoing mission is to ensure the preservation and protection of these sacred lands by getting the *Orans* registered in the revenue records, which would provide these places with a special status and thereby protection from getting exploited.

Importantly, the skepticism of local communities toward energy transition models labeled "green" or "sustainable" does not stem from opposition to developing new energy infrastructure for future progress. As Sumer Singh Bhati[59] asserts, "Development should be there, but we are against the destruction that is happening in the name of development, we don't want that (sic)." This raises grave concerns about "how it is brought into being"[60] and "how the mechanisms of transition are being operationalized"[61] whereby old extractive and exploitative practices colonize such spaces and energy transition fails to achieve its potential. On August 31, 2023, the government registered 12,588 *bighas*[62] of *Aalaji ka oran* land in revenue records. The land was reclassified from its previous status of *wasteland* and registered under The Rajasthan Land Revenue Act of 1956, under clause 92, which states:

> Land may be set apart for special purposes—(1) Subject to the general orders of the State Government, the Collector may set apart land for any special purpose, such as for free pasturage of cattle, for forest reserve, for development of *abadi* or for any other public or municipal purpose; and such land shall not be used otherwise than for such purpose without the previous sanction of the Collector.

Over the past 15 years, the local community has persistently advocated for the inclusion of *Aalaji ka Oran* in the appropriate revenue record category. Despite being rejected eight times, the registration was eventually approved on the ninth attempt. Preceding this accomplishment, on August 7, 2023, the government similarly documented approximately 6,000 bighas of land affiliated with *Degrai Oran* in Sanwta village, Jaisalmer. This action, though, was strategically timed just before the state assembly elections scheduled for November 2023, thus exhibiting how the registration of these *Orans* was not devoid of political motives. Such "localization of aeolian politics"[63] at these sacred spaces, by serving political agendas, demonstrates the ambivalences that still govern energy transitions in Global South/Majority world spaces, whereby solar and wind energy continue to be framed as transactional commodities.

Conclusion: Energizing the Humanities in Postcolonial Ecologies

In the multi-institutionally-authored 2015 article "Humanities for the Environment: A Manifesto for Research and Action," the contributing scholars ask, "What is the role of the humanities in the age of the Anthropocene?"—a query that is integral to understanding the ontologies of ambivalence in ecological transformations as centralized by this volume. Disciplines in the humanities and humanistic modes of inquiry, which offer an "understanding of human imagination, perception and relationship with their surrounding environments—both natural and social,"[64] provide possibilities and heuristics that are not bound within the limitations of instrumentalist approaches to renewable energy infrastructure. As the preceding sections highlight, the policies and politics of sustainable green energy in Global South/Majority World spaces cannot be perceived solely through dominant imaginaries of "common good," since "with weak global [environmental] politics, cultural distrust and imperfect communications, defectors are likely to get away with cheating."[65] The primary barriers that have emerged in many Global South/Majority World/postcolonial contexts are often seen in the inability of states, corporate institutions, and environmental stakeholders to acknowledge collective intelligence that may lie outside the purview of dominant and normative (Western) knowledge systems. Aeolian politics, in and around the *Orans* of Jaisalmer Rajasthan, provides a telling commentary, demonstrating that these community-owned lands are not only life-sustaining but also intrinsically connected to the dignity of the human inhabitants in these ecosystems. Such postcolonial ecologies foreground "a spatial imagination made possible by the *experience of place*,"[66] which is always based on lived realities that cannot be negated through temporal narratives of progress or enthusiasm for renewable energy that ignore "how human energetic desires—for light and heat, for movement and flourishing—[…] [can] deeply disrupt the energetic needs of other biotic life and systems on which we all depend."[67] Thus, the interdependent logics of environmentalism and politics of energy infrastructures need to be mediated with regional and cultural dialogue that is contextual and based within the everyday experiences of human inhabitants. These approaches may require us to move beyond current strategic goals that categorize climate crises within the register of "mega-problem[s] […] requiring mega solutions"—a paradigm that climate scientist Mike Hulme notes has led to a "political log-jam of gigantic proportions, one that is not only insoluble, but one that is perhaps beyond our comprehension."[68] In moving across a range of methodologies, including historical surveys, close reading, and humanistic fieldwork, where the "medium of encounter […] doubles the textual encounter with lived encounter,"[69] this intervention hopes to show how reflective analytical frameworks championed in the humanities can lead to contemplative research and co-produced action on key environmental challenges.

Notes

1 International Energy Agency, *World Energy Outlook 2021* (Paris: International Energy Agency, 2021), 3.

2 Suman Seth, "Putting Knowledge in Its Place: Science, Colonialism, and the Postcolonial," *Postcolonial Studies* 12, no. 4 (December 2009): 373–388, https://doi.org/10.1080/13688790903350633.

3 The Clean Development Mechanism (CDM) is a United Nations-run carbon offset scheme allowing countries to fund greenhouse gas emissions-reducing projects in other

countries and claim the saved emissions as part of their own efforts to meet international targets. It is one of the three Flexible Mechanisms defined in the Kyoto Protocol.

4 Viresh V. Patel, "Just Lip Service? Insights into the Sustainable Development Prospects of the CDM in Rural Gujarat, India," *Carbon Management* 4, no. 3 (2013): 291–307, https://doi.org/10.4155/cmt.13.22.

5 Boyer and Howe coin this term from the Spanish term for electricity derived from wind power in their duographs *Ecologics and Energopolitics* (Durham: Duke University Press, 2019).

6 Cymene Howe and Dominic Boyer, *Ecologics and Energopolitics: Joint Preface* (Durham, NC: Duke University Press, 2019).

7 Shahidul Alam, "Majority World: Challenging the West's Rhetoric of Democracy," *Amerasia Journal* 34, no. 1 (2008): 88–98, https://doi.org/10.17953/amer.34.1.l3176027k4q614v5.

8 Mohammad Arju, "ICCA Consortium," *ICCA Consortium*, February 27, 2021, www.iccaconsortium.org/2021/02/27/orans-rajasthan-india/.

9 Augé qtd. in Samuel Collins, "Head out on the highway: Anthropological encounters with the supermodern," *Postmodern culture* 7, no. 1 (1996): 9, our emphasis.

10 James Fairhead, Melissa Leach and Ian Scoones, "Green Grabbing: A New Appropriation of Nature?" *Journal of Peasant Studies* 39, no. 2 (April 2012): 237–261, https://doi.org/10.1080/03066150.2012.671770.

11 Alam, "Majority World," 88–98.

12 Deepak Sangroya and Jogendra Kumar Nayak, "Development of Wind Energy in India," *International Journal of Renewable Energy Research* 5, no. 1 (March 2015): 1–13, https://doi.org/10.20508/ijrer.v5i1.1626.g6466.

13 Our emphases; Tushar K. Moulik, "Energy and development options: the case of India," *The Journal of Energy and Development* 13, no. 2 (1988): 240, www.jstor.org/stable/24807776.

14 Moulik, "Energy and development options," 239–273.

15 Fuel produced from low grade coals.

16 Moulik, "Energy and development options," 239–273.

17 Moulik, "Energy and development options," 239–273.

18 Bernard Winraub, "India, Slow to Grasp Oil Crisis, Now Fears Severe Economic Loss," *The New York Times*, January 20, 1974, www.nytimes.com/1974/01/20/archives/india-slow-to-grasp-oil-crisis-now-fears-service-ecomomic-loss-a.html.

19 Moulik, "Energy and development options" (1988): 255.

20 The MNES, which was re-christened to its current avatar Ministry of New and Renewable Energy (MNRE) in 2006, has five key facets that make up its mission: 1. Energy Security, 2. Increase in the share of clean power, 3. Energy Availability and Access, 4. Energy Affordability, and 5. Energy Equity, with points 2, 3, and 4 having immediate considerations for aeolian politics in India.

21 The Hindu Bureau, "Right against Climate Change a Distinct Fundamental and Human Right, SC Judgment," *The Hindu*, April 8, 2024, sec. India, www.thehindu.com/news/national/right-against-climate-change-a-distinct-fundamental-and-human-right-sc-judgment/article68041693.ece.

22 The Hindu Bureau.

23 Howe and Boyer, *Ecologics and Energopolitics*, 194.

24 Ibid., 195.

25 Anupam Mishra, *Radiant Raindrops of Rajasthan* (New Delhi: Research Foundation for Science, Technology and Ecology, 2001), https://archive.org/details/TheRadiantRaindropsOfRajasthan/page/n1/mode/2up.

26 Nanditha Krishna and Aman Singh, *Ecological Tradition of India* (Chennai: C.P.R. Environmental Education Centre, 2014).
27 Nandita Singh and Om Prakash Singh, "Khadin: A Nature-Based Solution for Water for Agriculture in the Great Indian Desert," *Millenium Water Story*, October 15, 2018, https://millenniumwaterstory.org/Pages/Photostories/Water-and-Livelihood/Khadin-A-Nature-Based-Solution-for-Water-for-Agriculture-in-the-Great-Indian-Desert.html.
28 Marc Augé, *Non-Places: Introduction to an anthropology of supermodernity* (London: Verso, 1995).
29 Adani Renewables, "Building a Better Tomorrow Leading to a Future Powered by Renewable Energy," *Adani Green Energy Limited ESG Report 2022-23*, www.adanigreenenergy.com/-/media/Project/GreenEnergy/Sustainability/latest-report.pdf.
30 Nabodita Ganguly and J. Jayapriyanka, "Baring the Biodiverse Heart of the Thar Desert," *The Hindu Business Line* (August 5, 2023), www.thehindubusinessline.com/business-tech/baring-the-biodiverse-heart-of-the-thar-desert/article67163128.ece.
31 Kashish Shah, "IEEFA India: Renewables-Rich Rajasthan Has Potential to Lead India's Energy Transition," *IEEFA* (September 17, 2020), https://ieefa.org/articles/ieefa-india-renewables-rich-rajasthan-has-potential-lead-indias-energy-transition.
32 Judith Whitehead, "John Locke, Accumulation by Dispossession and the Governance of Colonial India," *Journal of Contemporary Asia* 42, no. 1 (February 2012): 1–21, https://doi.org/10.1080/00472336.2012.634638.
33 "[R]esource or land dispossession to further environmental or sustainability goals that has disrupted agrarian political economy and ruptured peasant moral economies", Ryan Stock, "Power for the Plantationocene: Solar Parks as the Colonial Form of an Energy Plantation," *The Journal of Peasant Studies* 50, no. 1 (September 2022): 162–184, https://doi.org/10.1080/03066150.2022.2120812.
34 John Locke, *Two Treatises of Government* (London: Whitmore and Fenn and C. Brown, 1821).
35 Stock, *Power for the Plantationocene*, 163.
36 Prakash Bhandari, "Sun Shines in the Thar Desert - Rajasthan Emerges as the Country's Biggest Solar Energy Hub," *National Herald,* December 25, 2021, www.nationalheraldindia.com/india/solar-energy-in-rajasthan-sun-shines-in-the-thar-desert
37 Aman Singh, "From stewards to trespassers: Pastoralist management of forest resources," *Pastoralist Livelihoods in Asian Drylands: Environment, Governance and Risk* (2017): 177–199.
38 Aman Singh, "From stewards to trespassers."
39 Jennifer Baka, "Making Space for Energy: Wasteland Development, Enclosures, and Energy Dispossessions," *Antipode* 49, no. 4 (February 2016): 977–996, https://doi.org/10.1111/anti.12219.
40 National Remote Sensing Centre, *Wastelands Atlas of India*, (Government of India, 2019), www.nrsc.gov.in/Atlas_Wastelands?language_content_entity=en.
41 Rajasthan Wind and Hybrid Energy Policy, 2019.
42 Ibid.
43 David Singh, " 'This Is All Waste': Emptying, Cleaning and Clearing Land for Renewable Energy Dispossession in Borderland India," *Contemporary South Asia* 30, no. 3 (July 2022): 402, https://doi.org/10.1080/09584935.2022.2099812.
44 David Singh, "This Is All Waste," 404, https://doi.org/10.1080/09584935.2022.2099812.
45 David Harvey, "The 'New' Imperialism: Accumulation by Dispossession," *Socialist Register* 40 (March 2004), https://socialistregister.com/index.php/srv/article/view/5811.
46 Dibyadyuti Roy, "Strategic Science vs. Tactical Storytelling: Disrupting Radioactive Masculinity through Postcolonial Ecologies," *South Asian Review* 37, no. 3 (2016): 37–61, https://doi.org/10.1080/02759527.2016.11978318.

47 Howe and Boyer, *Ecologics and Energopolitics,* 1.
48 Ibid., 6.
49 Rita Brara, "Are Grazing Lands 'Wastelands'? Some Evidence from Rajasthan," *Economic and Political Weekly* 27, no. 8 (1992): 411–418, www.jstor.org/stable/4397629.
50 Krishna and Singh, *Ecological Tradition of India.*
51 David McDermott Hughes, *Who Owns the Wind?* (London: Verso Books, 2021).
52 Hughes, *Who Owns the Wind? 28.*
53 Stock, "Power for the Plantationocene," 163.
54 David Singh, "This Is All Waste," 408.
55 This region covers the villages of Salkha, Kanoi, Chhatrel, Habur, Doojasar, and Kuchchari, i.e., sites for a project boasting a total capacity of 100.8 MW of wind energy and involving the installation of 48 Suzlon wind turbines. The expected annual energy supply to the NEWNE Grid of India is 187.8 GWh, resulting in a reduction of 178,948 tCO_2e emissions per year.
56 Also known as the "water wizard" of the Thar Desert, they have helped revive numerous traditional sources of water systems as well as *khadeen* using local ecological knowledge.
57 Nisha Paliwal, "The Problem with Wind Farms on the Sacred *Orans*," *Edge Effects*, May 23, 2023, https://edgeeffects.net/orans-wind-farms/.
58 Paliwal, "The Problem with Wind Farms."
59 A local activist from Sanwta village, Jaisalmer.
60 Howe and Boyer, *Ecologics and Energopolitics,* xii.
61 Ibid., 10.
62 A unit of land measurement that varies from region to region but roughly translates to 0.63 acres.
63 Howe and Boyer, *Ecologics and Energopolitics*, xiv.
64 Poul Holm et al., "Humanities for the Environment: A Manifesto for Research and Action," *Humanities* 4, no. 4 (2015): 977–992, https://doi.org/10.3390/h4040977.
65 Holm et al., "Humanities for the Environment," 979.
66 Our emphases, Elizabeth M. DeLoughrey and George B. Handley, *Postcolonial ecologies: Literatures of the environment* (Oxford: Oxford University Press, 2011), 4.
67 Howe and Boyer, *Ecologics and Energopolitics,* 2.
68 Mike Hulme, *Why we disagree about climate change* (Cambridge: Cambridge University Press, 2009), 332–333.
69 Shalini Puri and Debra A. Castillo, *Theorizing Fieldwork in the Humanities: Methods, Reflections, and Approaches to the Global South* (New York: Palgrave Macmillan, 2016).

9 Global South Ambivalences of Transformation?

Literature, Extractive Capitalism, and Literary Militancy in West Africa

Sule Emmanuel Egya

Introduction: Africa and the Ambivalences of Transformation

Africa is a continent endowed with rich oil and mineral resources, all of which attract Arabic, European, American, and now Chinese colonialists to the continent. I use the term "colonialists" in the context of this chapter, although they really do not call or present themselves as such but as industrialists interested in doing business with Africa. This disguise is at the heart of Global South ambivalences of transformation, because as industrialists they promise to provide socio-economic developments they actually have no intention of providing. Instead, they plunder eco-resources to develop and advance their own societies. Furthermore, while they talk of green transformations and associated ambivalences that occur when, for example, their newly built roads, water pipes, and temporary workers' settlements destroy traditional pathways, villages, and burial sites, most parts of the Global South, especially Africa, are faced with a lack of adequate and sustainable infrastructure. The desire for advancement in life in this part of the world is paradoxically truncated by the neglect and destructive impacts of extractive capitalism to the extent that local inhabitants, after failed attempts to press home their demands, resort to militant activities. The present condition, therefore, is that of visible eco-destruction matched by militant violence—as represented in Ogbowei's *Marshland and Other Poems* and Mbue's *How Beautiful We Were*. I therefore argue that the tensions thrown up by the gap between the desire for social-economic development and the failure of extractive industries to fulfill their promises of development pose a threat to notions of environmental transformation and sustainability that sit at the top of the Global North's agenda.

African nations, especially those with which I am concerned in this chapter, depend on the revenue generated from selling extracted resources for their existence. The narrative of extraction and cooperation appears to be smooth in the context of a nation's GDP and economic life, as efforts are made to focus on the profits and benefits of extractive capitalism for the nation, i.e., increase exports and provide more resources to build infrastructure. However, beneath this smooth surface often presented by investors, economists, and government officials, local populations have become wary of believing that these organizations will fulfill their promises to develop and bring modernity to the regions. In the context of oil

DOI: 10.4324/9781032627984-12

extraction in particular, initial hopes of progress and prosperity have given way to ambivalence and outright anxiety, and many locals now view oil-drilling activities as a curse instead of a blessing. Instead of adequate infrastructure, there are realities of diseases from pollution, gas flares, destructions of biodiverse life-forms, and the wiping out of entire communities.[1] This contradiction is underscored by crucial questions, such as: What real, concrete forms of development are enjoyed by local communities whose ecological wealth is being extracted? In what ways do socio-ecological, technological transformations, and modernized forms of life affect the ecological well-being—rich as it is with biodiversity—of these communities? And above all, how do these forms of transformation currently translate into a blessing for humans and nonhumans directly affected by the extractions?

These questions are asked, among others, by locals, activists, militants, and especially writers, the latter of whom are considered oracles by their societies. I am concerned here with the ways in which writers in Nigeria and Cameroon ask these questions in poetic and fictional forms, and how these questions translate into representations of the tensions and paradoxes arising from the gap between promises and fulfillment, between blessing and curse, which would be the kind of ambivalences of transformation experienced in most places in the Global South, where "extractive violence is [...] rather a core logic of the 21st-century global experience."[2] Nevertheless, more significantly, I am concerned with showing how writers go beyond ambivalence and advertising the plight of human and nonhuman victims of extractive capitalism to dramatize the militant ways in which local communities have fought back. I argue that given the severity of the curse (instead of a blessing) brought upon the communities by the activities of extractive capitalism, aligned with modernity's rather insurmountable machineries of eco-oppression, the writers tend to advocate a form of resistance that I would like to present as "literary militancy." This kind of resistance suggests being militant not only in the ways in which one is persistently alert to injustice, but also in how one embarks on guerrilla methods and insurgent manipulations to unleash terrors as the last resort to seek environmental justice. While this process is indeed ethically questionable, literary writers have felt compelled to dramatize and present it as a solution to the persistence of eco-destruction that extractive capitalism unleashes on unsuspecting communities, with the justification being that capitalists deploy even worse violent methods, even if disguised. I predicate this concept of literary militancy on the ideas generated by both protest literature in Africa and postcolonial ecocriticism. I use G. Ebinyo Ogbowei's poem "Marsh Boy" and Mbolo Mbue's novel *How Beautiful We Were* as illustrations in this regard in the context of oil extractions in Nigeria and Cameroon. I make the point that these texts do not represent ambivalences of transformation in the context of the Global North, where technological transformations inevitably result in tensions between those who embrace it and those who do not. In the case of Africa, as captured in these texts, there is nothing like ambivalence as a reaction to technological transformations because the technology does not even exist. The question is therefore not about technological transformations and their ambivalences but about confronting existential crises of survival unleashed by extractive violence. Before presenting a brief analysis of the texts,

I attempt in the next section to provide the socio-historical and political context of extractive violence and to show how literary militancy has emerged as a form of aesthetics of resistance.

Oil Extraction, Consequences, and the Rise of Literary Militancy

The extraction of oil and other resources in Africa began very soon after the colonial invasion. It is worth recalling that the two-pronged purpose of the colonial expansion of empire in Africa, on the one hand, was to civilize "primitive' peoples" and, on the other hand, to take away the wealth of the continent.[3] These factors rooted the Christian mission, the transatlantic slave trade, and political partition. In this context, the history of extractions began with the slave trade, the removal of artifacts and cultural objects from Africa to the Middle East and Europe, and the early trade in resources such as rubber, copper, palm oil, and cocoa under very exploitative terms, thereby systematically dispossessing locals and relocating whatever they found useful to their homelands. These campaigns against the continent resulted in acts of resistance by some African rulers and community chiefs, which the colonialists severely dealt with in the early 20th century. For instance, in Nigeria, bitter struggles and military operations led to the deposing and exiling of leaders such as Oba Ovonramwen of the Benin Kingdom, Nana of Itshekiri, and Jaja of Opobo.[4]

Oil extraction became a chapter in the history of colonial expansion in Africa at a time when fossil fuels were being discovered as sources of energy across the world. West Africa is particularly blessed with fossil fuels, and countries such as Nigeria, Cameroon, and Equatorial Guinea have large reserves of oil and gas.[5] The first commercial discovery made in Nigeria was in 1957 at Oloibiri, in the present Bayelsa State, while in Cameroon, reserves were first discovered in 1977 in the Rio Del Rey basin. Prospecting and extraction operations were carried out by foreign corporations such as Perenco Oil and Gas, Total, and Shell BP, with the countries' former colonial authorities having great stakes in these organizations. Today, Nigeria and Cameroon receive a huge amount of foreign income from fossil resources such as crude oil and gas, which together constitute the nations' biggest exports. Paradoxically, products refined from oil, mainly PMS (premium motor spirits) and diesel, account for the largest imports, but the two countries cannot refine even half of the crude oil they need and are unable to put in place functional refineries to cater for their home energy needs. Moreover, there are no visible developments in these zones of extraction. The situation is compounded by the destructive effects of extractive corporations that result in air and water pollution, gas flaring, oil spillage, and the destruction of biodiversity. Moreover, humans suffer from mysterious diseases and die slowly because of the water they consume and the air they breathe in communities where oil corporations are located. This situation is well represented in *How Beautiful We Were*, which details a deadly coughing disease that hits vulnerable children particularly. This destruction of the human body in this form instills a sense of urgency to violently confront the eco-oppressor (in this case, oil corporations and governments). In this context, ambivalences that are often mentioned as unpleasant yet necessary side-effects of advanced developments, in whatever form, become meaningless, as survival instincts take center stage.

It is usually locals who initially act against extractive capitalism, for the obvious reason that they are the worst hit, physically, psychologically, and spiritually. Resistance is often in the form of group activities whereby some people, mostly the youth, come together to create unity and awareness among locals with the aim of confronting oil corporations located in their communities. It is a gradual process in which these groups seek dialogue with the corporations and the government, present their grievances, and demand compensation and due reparation for environmental damage. When this dialogue fails—and it often does—the groups resort to aggression, sometimes spilling over into full violence, to pressurize corporations and governments to accede to their demands.[6] In some cases, the locals are brought to the table for dialogue with their oppressors, but such meetings turn out to be a ruse, as we see in *How Beautiful We Were*.[7] And so aggression and violence only grow worse, as most of the time corporations and governments prefer to use force through the military to counter attacks made by local groups. Water sources, the air, trees, rocks, and animals suffer a worse fate in these conflict zones.[8]

Modern-day environmental activism in West Africa may be tied to the pioneering activities of the writer and activist Ken Saro-Wiwa, who hails from the oil-rich Niger Delta region of Nigeria. Scholars such as Rob Nixon, Byron Caminero-Santangelo, and Stephanie Newell have variously commented on the worldwide influence of Saro-Wiwa as an environmental activist. Nixon frames him—along with the Kenyan environmental activist Wangari Maathai—as a "writer-activist" and posits that such individuals can "help us apprehend threats imaginatively that remain imperceptible to the senses."[9] The tragedy surrounding Saro-Wiwa's ultimate demise continues to be narrated and inspire literary and political activism,[10] which is so strong that current writers in Nigeria continue to portray him as a character in literary works that historicize eco-oppression, in what I have referred to in an earlier work as the "Saro-Wiwa imaginary."[11]

Saro-Wiwa's struggles represent the stage of dialogue in environmental activism in Nigeria. He founded the Movement for the Survival of the Ogoni People (MOSOP) to converse with oil corporations and the government regarding the misery facing the indigenous people of Ogoniland. Rather than engage in conversations with MOSOP, the government and corporations criminalized his activities, arrested him, hastily tried him in a military tribunal, and executed him by hanging. The killing provoked an outburst of violent activities, framed as "militant struggles," across the Niger Delta, seen in the form of youth unrest, guerrilla tactics, and armed conflicts that multiplied in the region, producing gang leaders in the form of militant warlords.[12] More importantly for this study, the judicial killing of Saro-Wiwa also provoked an efflorescence of environmental writing in the forms of poetry, fiction, and drama that historicized the reality of failed development. This literature insists on the irrelevance of ambivalences resulting from infrastructure development since the communities have not actually witnessed any substantial development. It also creates awareness and raises the consciousness of readers against the government and oil corporations responsible for multiple eco-crises.

Beyond Saro-Wiwa, current struggles in the Nigerian oil region are militant and do not shy away from open acts of violence. Militancy, in this context, may be seen as an extreme form of activism, i.e., one that locals resort to in order to seek environmental justice because attempts at dialogue have failed. While there are militant groups in Nigeria and in Cameroon, using aggressive means to press home their demands regarding environmental justice, there are also authors who write by using what one might describe as *violent* language to arouse the consciousness of readers against agents of eco-oppression. The process of producing such literature—its conception, textual formation, and reception—is what I describe as "literary militancy." In most cases, these writers originate from those communities facing eco-destruction. Many of them have observed the processes involved in environmental degradation from early childhood and have been denied basic things in life such as hospitals, schools, roads, drinking water, and electricity, among others, not only because of the pollution around them, but also as a result of the total neglect of the zone after it has been abandoned by extraction businesses.[13] In this regard, they are therefore highly alert to the inhumanity unleashed on their communities, and they depict it in the best way possible by reminding readers that, for a multiplicity of reasons, not all members of a community can be expected to move elsewhere. Most importantly, however, they use what the Gambian poet Tijan M. Sallah calls "the magical potency of language"[14] to disparage eco-oppressors.

The notion of literary militancy is anchored in African protest literature, which began during the pioneering period in which writers intentionally embarked on a counter-discourse to reverse colonial narratives. Chinua Achebe, one of the first in this regard, aptly frames the intentional counter-discourse when he says "I would be quite satisfied if my novels [...] did no more than teach my readers that their past [...] was not one long night of savagery from which the first Europeans acting on God's behalf delivered them."[15] The protest writing of the Achebe generation is followed by that of the second generation of African writing, which focuses on lamenting post-independence disillusionment and attacking homegrown dictators who inherit what Achille Mbembe calls "the structure of commandment"[16] from colonial authorities and practice it to ridiculous levels. This is illustrated in the works of, among others, the Malawian poet Jack Mapanje, the Ghanaian poet Kofi Anyidoho, and the Nigerian poet Niyi Osundare. It is in this tradition of protest writing, namely, of using literature as an instrument to change the condition of existence, that I locate the literary militancy canon.

Literary militancy is connected to the aspect of postcolonial ecocriticism that deals with environmental justice. Postcolonial ecocriticism, as Graham Huggan and Helen Tiffin suggest, is more about "political processes"[17] involved in changing the world that postcolonial subjects inhabit. That is, it begins from the premise that postcolonial nations such as Nigeria and Cameroon face ecological crises that are underscored by neo-colonial forces, and any attention given to these crises must first confront transnational extractive corporations that operate in colonial modes. Literary militants, such as Ogbowei and Mbue, are aware of the colonialism that profoundly shapes extractive capitalism, and their quest for environmental justice is indeed a pursuit to seek political, cultural, and ecological decolonization. As Caminero-Santangelo writes, "in terms of environmental representations, [African] writers are concerned with lived environments, the social implications of environmental change, and the relationships between representations of nature and power."[18] (13). It thus transpires that what African literary militants ultimately desire is socio-ecological justice based on political, social, and ecological emancipation.

"Marsh Boy:" A Militant Eco-Poetics

Ogbowei is a native of Bayelsa State, one of the territories in the Niger Delta region of Nigeria. He grew up in the region, and, although he has traveled to other parts of Nigeria mainly for educational purposes, he has lived most of his life in Bayelsa. This background is fundamental to his writing career, since he writes poetry about the region with a sense of belonging often implied in the we-versus-them dialectic that runs through his eco-poetics. By his "eco-poetics," I mean his aestheticization of the condition of eco-destruction in his birthplace, the ambivalences of failed development associated with extractive capitalism, and his people's experience of environmental injustice. His eco-poetics sharply contrast the idea of the Global North ambivalences of transformation, in that he poeticizes the pains of systematic underdevelopment. Ogbowei has consistently resolved, on behalf of the people of his oil-rich birthplace, to attack extractive capitalism—as sustained in his collections, which include *The Heedless Ballot Box* (2005), *Let the Honey Run and Other Poems* (2013), *Marsh Boy and Other Poems* (2013), and *Matilda* (2018).

"Marsh Boy", the lead poem in his collection *Marsh Boys and Other Poems*, illustrates Ogbowei's militant eco-poetics in two ways. First, it eulogizes the real acts of militant struggle in the region, which became pervasive in the 1990s after the judicial killing of Saro-Wiwa. What many saw as the state murder of the environmental activist marked the end of dialogue between activists and oil capitalists and signaled the beginning of violent anti-extraction campaigns. Second, it demonstrates Ogbowei's use of *violent* language against extractive capitalism and the institutional powers behind it, and this aligns with the gospel of militant struggles in the Niger Delta anchored by achieving environmental justice by any means, even if violent. Evidently, this stands against the Global North rhetoric of ambivalences that downplays the straightforward eco-violence of extractive industries. "Marsh Boy" is a long poem of over one hundred and fifty lines that depicts a militant's righteous outrage and justification for taking up arms—a case of a long-suppressed subject exploding with uncontrollable energies to right the wrongs done to his society. The first stanza is unmistakable about this cause:

> locked in the prison of poverty
> denied the right to rise
> out of holes in sighing swamps
> i sing of creeks with crushed dreams

i sing of ponds with decaying hope
i sing of rivers the cargo ships and
                supertankers sail on
i sing of swamps sold to swindlers and
                rustlers
i sing of swamps demanding dignity
                and opportunity.[19]

The lines above fly in the face of what has been framed as "ambivalences of ecological transformations," as the persona emphasizes the rawness and urgency of his mission, i.e., to confront extractive capitalism even if their super tankers ("supertankers") run on clean energies. The first three lines set out the rationale: not only the reason for resistance but also the reality that the opportunity for dialogue is foreclosed. The persona likens his (let us assume that he is a male youth) existence to a "prison of poverty," implying that institutional powers have created a condition of imprisonment not only for himself but for the entire human and non-human lives in his community, which he chooses to refer to as "swamps." Being in prison amounts to denying one the right to live one's life in the way they choose to live it. They are swamp people (that is, they live in lowlands mostly covered with waters) who would normally live their lives fishing and farming—a way of life that oil extraction has denied them. From the denial, which is the first sign of eco-oppression, the marsh boy (notice how the name "marsh boy" reflects the locality of his community, which is in a marshland) decides to sing. Singing here is a metaphor for complaining about eco-destruction and for celebrating militancy; on the one hand, it exposes the incidents of eco-oppression the community faces and, on the other hand, it eulogizes those engaged in militant struggle. Indeed, there is a great deal of self-praise in the poem, in which the marsh boy presents himself as the gun-toting messiah of his eco-oppressed communities. He sings of ponds that are decaying, of rivers burdened with "cargo ships" and "supertankers," i.e., the vehicles used to convey the mined wealth from his community. In this regard, he points out, his community has been "sold to swindlers and/rustlers." He also sings of "ambushed agreements," which suggests that any agreements that have been reached between his community and extractive corporations—a practice that usually takes place at the beginning of extractive activities to build trust—have not been kept.

In the second stanza of the poem, we see the marsh boy not as a singer but as a gun-toting militant, given the failure of agreements. He is no longer asking for dialogue, seeking the attention of the institutional powers, but has now decided to embark on extrajudicial actions that he feels will draw attention to his plight and, possibly, give him justice. He presents himself as follows:

i am the marsh boy
        quick and handy with a gun
i am the marsh tiger
        stalking beneficent tyrants.[20]

This reflects the reality of militancy in the Niger Delta region, where the youth decided in the 1990s to take up arms to confront oil corporations. In what turned out to be prolonged violent conflicts marked by kidnapping, maiming, and killing, the government started a military operation known as the Joint Task Force (JTF) to protect the oil workers and their installations and to fight the militant youth.[21] Ogbowei, like most poets from the region, celebrates the militant youth in this poem, clearly taking a position that supports their violent activities. The tone of celebration runs throughout the poem as the poet presents the marsh boy as a messianic hero who will liberate the community from a "prison of poverty." Furthermore, the boy is "the dagger/driven into the mind of mugger,"[22] i.e., oil capitalists. Additionally, he is the "bomb/exploding the peace of the pillage,"[23] as oil corporations, rather than bringing the development they promised the community in the beginning, have stolen the people's natural resources. Notice the use of the violent words "mugger" and "pillager" (coined from "pillage") as metaphors for the eco-oppressor. Further down, the poem describes the activities of militant youth as the condition in which anger and outrage reign supreme, since reason—the rational way of dealing with conflicts arising from extractive activities—has been abandoned. The poet likens militancy to "angry waves" that "wash away contentious coasts," "wash away pumpkin kings," and "wash away turkey cocks,"[24] all of which are metaphors referring to the intuitional powers that are responsible for the Global South ambivalences of transformation in the region, marked by their intentional failure to keep to their promises of development. It is against their treachery that the youth embark on violent actions, what the poem, using another metaphor, describes as "vengeful cutlasses and foxy axes" that "hack off the grabbing hands of a dying dominion."[25] "Vengeful" here suggests that the militants, indeed the communities, believe that their retaliation is justifiable.

Ogbowei's militant eco-poetics represents a kind of literary response to extractive capitalism in Nigeria. The response, as the poem demonstrates, is negative, most times hostile, because extractive capitalism in Nigeria, as evidence shows and most analysts agree, has not brought the desired socio-economic development to the nation. Worse, it has clearly caused serious ecological damage in the communities where oil extraction takes place, most of which are helpless, in that the government fails to protect them because of their connivance with the oil corporations. Some of the government officials and public servants that benefit from the corrupt ties between the administration and oil corporations are natives of the oil-producing communities; they are also local chiefs involved in corrupt practices and are caught up in what the poem describes as "the bizarre rituals of corruption."[26] The poem asks a rhetorical question about them:

> how can they live in peace
> whose seductive symphony
> calls out of ravaged rivers
> the avenging demons of destitution.[27]

In other words, oil capitalists, government officials, local chiefs, and other powerful people who enrich themselves while the communities suffer in poverty cannot live in peace because they have created "demons of destitution" that must fight them. The persona "marsh boy" represents these demons and fights back on behalf of the community. Like "marsh boy", Ogbowei the poet sees himself as an oracle speaking on behalf of his community. In the preface to the collection, he emphasizes that "A messenger, chronicler and entertainer [which is what he considers himself] who speaks the mind of his people, his message isn't his, but that of his community."[28] Nevertheless, he is a messenger who is as militant as the marsh boy. "Ogbowei's articulation," G. G. Darah writes in his foreword to the collection, "is unique because he is not content with lamentation and denunciation; he salutes and celebrates the heroism of those who recognize the quality of freedom that issues from the barrel of the gun."[29] And with his choice of diction, which is as violent as the gun, Ogbowei ranks as one of a number of militant poets from the Niger Delta region of Nigeria.

How Beautiful We Were: Oil, Modernity, and the Fate of Kosawa

A Cameroonian by birth but now living in the United States, Mbue offers a rather protean x-ray of the diverse effects of what has been called "petrodollar capitalism" on indigenous communities in Africa. Although her country, Cameroon, is also oil-producing and faces all the environmental dangers of oil extraction despite its relatively small output, Mbue appears to be more concerned about the fate of Africa and its indigenous inhabitants generally. The specifics of Cameroon, as we know it today, are there in the novel: a country ruled for a long time by a civilian dictator, self-conflicted by the French-English dichotomy of its population, totally weakened by neocolonial forces. But these are also the specifics of many African countries, including Nigeria. Moreover, the novel is clearly preoccupied with failed development in extractive zones across Africa that destabilizes the notions of Global North ambivalences of transformation.

Whereas Ogbowei's poetry, as we have seen above, uses the metaphor of the "marsh boy" to demonstrate the point at which dialogue collapses and is replaced by a violent campaign, Mbue's novel gives us a total spectrum of the struggle. That is, *How Beautiful We Were* dramatizes how the people are persuaded that oil extraction brings blessings but are rudely awakened to its destructive effects, how they seek dialogue with the oil corporation but are deceived, and how, totally frustrated, they resort to violent activities. This is what constitutes Global South ambivalences of failed development, in contradistinction to the ambivalences of transformation in the Global North. It is a shift to literary militancy rather than debates about ambivalences in a technologically transformed environment. The novel further dramatizes the aftermath of militant activities and depicts the tragic end of a community totally crushed by modernity. That Kosawa village, despite all efforts, including the militant actions of its youth, is turned into oil wells is a way of erasing its existence at the end of the novel and suggests the weight of capitalist modernity as a system of beliefs and norms that supports technological rationalization, the exploitation of nature, and the eradication of traditional culture. As the novel shows, extractive capitalism is a very potent instrument of modernity's destructive power. After the long battle with the multinational oil company Pexton, which climaxes in the kidnapping of the organization's overseer Mr Fish and his wife, Kosawa succumbs to the final blow of the violence modernity unleashes on local communities. Kosawa must give way:

> We were refused one last chance to enter Kosawa and empty our huts. The government decided the land had become too contaminated for human presence. His Excellency ordered Kosawa burned. What once were our huts became ashes. Our mothers' kitchens, ashes. Our barns and outhouses, ashes. Our ancestors' pride, ashes. Nothing remained of Kosawa, except for what we kept in our hearts.[xxx]

What starts with the promise of new development, as oil was initially being drilled, turns out to be the gradual erasing of the community as the people struggle in vain to keep their land and their lives. At every point, Pexton presents a façade of being a corporation that cares about the people living on the land it drills, and it pretends to listen to their grievances and even tell them lies while conniving with governments in Africa and in America to ensure that nothing stands in its way of drilling to death the land of Kosawa. Even at the point of the final salvo, when they start drilling on the place where Kosawa once stood, Pexton's wily act of extermination is demonstrated in the following sarcasm: "Pexton created a scholarship in honor of the overseer [Mr Fish] and his wife, the Augustine and Evelyn Fish Memorial Peace and Prosperity Scholarship. The scholarship was for [Kosawa's] children alone[…]. They'd already begun digging a new well in what used to be our village square when they announced scholarship."[30] They send the children to America to get educated and give them jobs over there so that they can forget about their village. It is the ultimate assimilation of local communities, peoples, and cultures by modernity through petrodollar capitalism. Perhaps the novel's greatest message in this regard is the following paradox: "Sometimes we ask our children about the cars they drive. The cars seem to be bigger than they've ever been, needing more oil. Do they think about it, about the children who will suffer as we once did just so they can have all the oil they want?"[31] It is ironical that the children of Kosawa now have scholarships from Pexton, the oil corporation that turns their village into oil wells, with which they travel to America to study, work, and live big by driving big cars. Would this be regarded as the transformation, the development that extractive capitalism brings to the community? It is not, and that is why Thula rejects it and returns to Cameroon to continue the struggle. She can understand the deceit of giving them an education and jobs in the US while simultaneously wiping out their biodiverse eco-existence in Kosawa. It is a function of destructive modernity that the new generation of Kosawa villagers is not interested in how drilling oil has led to the death of Kosawan children, the brutality, rape, and murder visited on Kosawa in the past—that they may not care that their ancestral land has been wiped from existence.

Mbue's literary militancy counters such devastating modernity, which she does by persistently staging the defiant voice of the community, collectivized in the pronoun "we"—a voice that refuses to die even when Kosawa village is destroyed. It is the voice we hear at the beginning of the novel: "We should have known the end was near. How could we not have known? When the sky began to pour acid and rivers began to turn green, we should have known that our land would soon be dead."[32] It is also the voice we hear at the end of the novel: "It may be long dead, Kosawa, but we never forget it, the splendid piece of the earth it was. We can never forget it, for there our spirits were whole."[33] They tell the story of Kosawa, but more importantly, they tell their children the story of Thula, the fiery activist of Kosawa. In these tales, the culture, folklore, and the beautiful ecology of Kosawa live on. In this regard, storytelling, as it was in the past, becomes the means of asserting one's existence and heritage. In traditional African society, knowledge, systems, and folklore are passed from one generation to another through storytelling.[34] As long as there are people to tell stories of the past, a community, an ethnic nation, or a people never dies. Kosawa therefore resorts to this pre-modern way of self-preservation through storytelling, in order to survive the onslaught of petrodollar modernity.

It is useful, in the frame of literary militancy, not to think of *How Beautiful We Were* as a story of defeat but as one of eco-oppression and eco-destruction brought upon a violated community by petrodollar corporate liars and murderers whose modus operandi best exemplifies the end of the ambivalence of transformation. It is also a story of people who embark on militant violence as a last resort to fight for their existence, their right to live in their native land, after all avenues of peaceful negotiations have failed. When Pexton first arrives in Kosawa, it brings with it a promise of development, the notion that life will improve for the community if they welcome oil-drilling on their soil: "The [oil] representatives told [the villagers] that drilling for oil would bring something called "civilization" to [their] village. One day, the government representatives said, Kosawa would have a wonderful thing called "prosperity"."[35] And yet the villagers face virulent diseases that kill children, their water becomes unfit for drinking, and their soil becomes incapable of growing crops. Any move they make to get Pexton's attention is resisted, and even when the new Pexton helmsman Mr. Fish eventually listens to them and gives them promises, nothing useful comes out of their engagement. *How Beautiful We Were* is also a story revolving around a militant campaign. The evolution of a group of some aggressive youths, known as The Five, who take their violent struggle to Pexton, marks the inevitable turn in all battles against extractive capitalism, and all of those against the ambivalence of development, marked by deceit in oil-drilling zones, must necessarily end in violence, albeit it is usually the last resort, as the novel clearly demonstrates. In spite of her zeal to see her community free of the diseases associated with drilling for oil, Thula initially tries as best she can to avoid violence. However, she gradually becomes persuaded to give the youths money to purchase arms and ammunition when she is increasingly faced with reality. She insists that she has to give authorization before the weapons are used. That is not to be, though, because in most cases, the eventual outburst of violence is spontaneous. This is how the novel depicts it:

> Our guns might have remained hidden in the forest forever, awaiting Thula's blessing for us to use them, if one of us hadn't returned from visiting a relative in another village to find his son dead[…]. That evening, after the funeral, our friend decided his wait was over[…]. He dug out his gun and headed into the bushes at the edge of the Gardens. From behind a tree, through his gun's telescopic sight, he observed three laborers smoking pipes[…]. He killed them all.[36]

This marks the beginning of a violent campaign, which culminates in the kidnapping of the chief executive officer of Pexton and his wife. Most struggles against extractive capitalism in West Africa end with such violence, which is well represented in most environmental writings, especially from Nigeria.

Mbue's narrative, however, is unique in the sense that—more than any novels before it (such as Helon Habila's *Oil on Water*, Tanure ojaide's *The Activist*)—it depicts a sustained game of violence that dramatizes the limit of the agency of the local community in the face of vicious modernity. On the surface, it turns out that the government and the oil corporation are the agents of extractive capitalism, powerful enough to destroy Thula's years of struggle in no time. Progressively, all efforts, both peaceful and violent, made by Kosawa's children to keep their ancestral land appear to be in vain. The novel's emphasis, which some readers may find troubling, remains that extractive capitalism is so powerful and vicious that no level of countermeasures can stand in its way. While it appears that the novel undercuts its own goal by reducing Thula and her friends' militant struggles to ashes, as their deaths suggest, a careful reader will see that it is possible for another Thula to emerge from among a new generation who love to listen to the stories of Kosawa, for the struggles and legacies of the militant youth will remain alive in them.

It is worth noting that Mbue may not have presented victorious militant characters in the same way that Ogbowei does, but she nevertheless deploys similar violent language (note the satiric punches against Woja Beki and His Excellency, for instance) against the eco-oppressors and an equal amount of empathy for the eco-victims. The narrative reveals the vicious undertones of extractive capitalism, including support for a neo-colonially backed dictatorship that politically, socially, and culturally emasculates the citizenry. Beyond raising her readers' consciousness about the crushing of an indigenous community by a powerful oil corporation and its allies, Mbue as a militant writer inspires Cameroonians—indeed Africans—to rise up and fight for socio-ecological justice.

Conclusion

I have attempted to show that there is a growing response of writers to petrodollar capitalism's Global South ambivalences of development, characterized by deceit and failed promises. This response mainly takes the form of literary militancy, which is rooted in the ideology of African protest writing. Started as a response to colonialism, African literature remains an aesthetic reaction to modern and neo-colonial forces that confront the continent today, such as using eco-resources from extractive zones in the Global South to fund green transformations in the Global North. Extractive capitalism in Africa implies dispossessing locals, and this not only stalls the continent's development, but also unleashes systematic violence aimed at wiping away the existence of humans and nonhumans, as Ogbowei's poetry and Mbue's narrative depict. Literary militancy, i.e., the type we see in the works of these writers, implies that African writers, who traditionally figure as oracles of their society, must remain vigilant—indeed militant—to ensure that indigenous communities in Africa are not eventually wiped out by petrodollar modernity.

Notes

1 It interesting that the Nigerian poet Nnimmo Bassey, in direct response to the idea of an oil curse instead of an oil blessing, titles one of his poetry collections *We Thought It Was Oil but It Was Blood* (Ibadan: Kraft Books, 2002).
2 Judith Shapiro and John-Andrew McNiesh, introduction to *Our Extractive Age: Expressions of Violence and Resistance* (London: Routledge, 2021).
3 See Frantz Fanon, *The Wretched of the Earth* (New York City: Grove Press, 2004).
4 See Ike Okonto and Oronto Douglas, *Where Vultures Feast: Shell, Human Rights, and Oil* (London: Verso, 2003).
5 Nicholas Shaxson, *Poisoned Wells. The Dirty Politics of African Oil* (London: Palgrave, 2007).
6 Okonto and Douglas, *Where Vultures Feast.*
7 In the Niger Delta region, for instance, the case of Ogoni is instructive. They began a dialogue with governments and Shell BP in the 1990s, but none of their demands to date regarding cleaning the environment have [SE: does this plural form go with "none" which means "not any one of"?] been fully met. See the report "Ogoni cleanup. Federal govt. gives reasons for slow pace at work," *Premium Times* (December 22, 2022) about the delayed cleaning of the Ogoniland: www.premiumtimesng.com/news/top-news/571745-ogoni-cleanup-federal-govt-gives-reasons-for-slow-pace-of-work.html
8 See Cajetan Iheka, *Naturalizing Africa: Ecological Violence, Agency, and Postcolonial Resistance in African Literature* (Cambridge: Cambridge University Press, 2018).
9 Rob Nixon, *Slow Violence and the Environmentalism of the Poor* (Cambridge: Harvard University Press, 2011), 15.
10 Stephanie Newell, "Introduction: Ken Saro-Wiwa as a Public Intellectual," *Research in African Literatures* 14, no. 4 (2017), vii-xvi.
11 Sule E. Egya, "Literature, Activism, and the Saro-Wiwa Imaginary," in *The Literature and Arts of the Niger Delta*, eds. Tanure Ojaide and Enajite Eseoghene Ojaruega (London: Routledge, 2021), 243–253.
12 See Okonta and Douglas, *Where Vultures Feast.*
13 Robert O. Dode, "The Political Economy of Resource Curse and the Niger Delta Crisis in Nigeria: Matters Arising," *Niger Delta Research Digest*, 5 (2011), 96–105.
14 Tijjan Sallah, "Introduction," in *New Poets of West Africa*, ed. Tijjan Sallah (Lagos: Malthouse Press, 1995), 16.
15 Chinua Achebe, "The Novelist as Teacher," in *African Literature: An Anthology of Criticism and Theory*, eds. Tejumola Olaniyan and Ato Quayson (Malden: Blackwell Publishing, 2007), 105.
16 Achille Mbembe, *On the Postcolony* (Berkeley and Los Angeles: California University Press, 2001), 42.

17 Graham Huggan and Helen Tiffin, *Postcolonial Ecocriticism: Literature, Animals, Environment* (London: Routledge, 2010), 13.
18 Caminero-Santangelo, Byron, *Different Shades of Green. African Literature, Environmental Justice, and Political Ecology* (Charlottesville: University of Virginia Press, 2014), 13.
19 [SE: G.] Ebinyo Ogobwei, *Marsh Boys and Other Poems* (Ibadan: Kraft Books, 2013), 21.
20 Ogbowei, *Marsh Boys*, 21.
21 Okonta and Douglas, *Where Vultures Feast.*
22 Ogbowei, *Marsh Boys*, 22.
23 Ibid., 22.
24 Ibid., 22–23.
25 Ibid., 23.
26 Ibid., 25
27 Ibid., 25.
28 Ibid., 7.
29 G. G. Darah, foreword to *Marsh Boys and Other Poems,* by G'Ebinyo Ogbowei (Ibadan: Kraft Books, 2013), 12–13.
30 Mbue, *How Beautiful*, 252.
31 Ibid., 252.
32 Ibid., 6.
33 Ibid., 253.
34 See Isidore Okpewho, *African Oral Literature: Background, Character, and Continuity* (Bloomington and Indianapolis: Indiana University Press, 2009).
35 Mbue, *How Beautiful*, 54.
36 Ibid., 213.

10 Mining for a Low-Carbon Economy?

Articulations of the Mexican Corporate Sector

Rafael Hernández Westpfahl

Introduction

Green transformation processes imply ambivalences—as demonstrated by global efforts toward decarbonization. In the last few years, the World Bank has predicted a long-term boost in demand for minerals required for technologies associated with a global energy transition toward a low-carbon economy.[1] Meanwhile, these "green" minerals, as core components of renewable energy technologies, are inherently ambivalent. On the one hand, they play a crucial role in mitigating greenhouse gases by enabling a gradual switch from fossil fuel-based energy provision to so-called "clean" technologies. On the other hand, they are harvested via an intensive extraction process accompanied by high environmental and social costs.[2]

In response to the high demand for these minerals, assessments show a worldwide rise in mining exploration. Latin America represents probably the most important region for such activities, with Mexico being the country with the highest number of new mining explorations in the region. While the capacities of big lithium reserves in the federal state of Sonora dominate debates on mining,[3] more importantly, Mexico ranks within the top ten countries extracting 17 minerals, among which are silver, copper, and zinc.[4] Silver is considered one of the most essential metals necessary for technologies required for a so-called "green energy transition," with an expected increase in demand of 56% by 2050.[5] Photovoltaic solar panels, electric vehicle batteries, and corresponding charging infrastructure all need silver.[6] Nonetheless, research in and on Mexico's mining sector, against the backdrop of the global energy transition beyond the recent emphasis on lithium,[7] has so far been understudied.

Research in and on Latin America and beyond is currently interested in and increasingly concerned with questions relating to how different social actors provoke, accelerate, slow down, perceive, and reflect on crises and transformation processes.[8] A growing body of literature problematizes alternative forms of extractivism in which control over land and resources is reproduced in the names of environmental protection, climate change mitigation, and energy transition.[9] Backhouse and Lehmann,[10] for instance, add to the debate on new frontiers of land control,[11] referred to as the "renewable frontier," while others[12] have investigated this concern in the so-called "Lithium Triangle" between Argentina, Bolivia, and

DOI: 10.4324/9781032627984-13

Chile. These contributions underscore the central role played by discourse strategies in "green extractivism." In this light, mining conflicts are not only struggles over territories and their resources, but the dimensions of perception and knowledge of such projects are also contested. Coming back to the initial concern regarding how social actors perceive and reflect on transformation processes, a crucial matter is the strategies these actors adopt and how they are articulated. Hence, in this contribution, I ask what discourse strategies are employed by mining companies in Mexico to promote acceptance of mining?

By analyzing public documents published by mining companies operating in Mexico, this contribution aims to extrapolate the different discourse strategies used to obscure the social and ecological costs of mineral production. I position this account within the broader discussion on the ambivalences of ecological transformations. For this purpose, the chapter is structured as follows. First, in the conceptual framework, I draw on the operationalization of discourse strategies in the context of new land control frontiers, especially on the articulation concept derived for this purpose from Stuart Hall. Subsequently, methodological considerations are briefly pointed out before the empirical analysis of the examined documents from mining companies in Mexico is presented. Thereafter, the analyzed results are compared and discussed. In the conclusion, I take up the question posed in the introduction and reflect on the limitations and reach of the paper.

"Green" Mining as a New Frontier

Theoretical contributions in the field of political and social ecology point to the embeddedness of nature in social relationships, which by themselves are dialectically constituted by discourses. Thus, discourse is fundamental to knowledge on nature. Understood in this way, social actors produce and transform nature, often with different interests and unequal power resources.[13] Focusing on discourse as a conscious and directed effort to produce knowledge implies an understanding of knowledge production as a power-laden process. The decisive role discourses play in mining conflicts has long been underscored in political ecology literature.[14] As well as struggles over territories and their resources, they are also, in a broader sense, "contests over knowledges, worldviews, identities and the configuration of development."[15] Within territorial conflicts, discourses are strategically applied by all actors. Corporate actors, for example, whitewash the socio-environmental harm done by their resource extraction through environmental discourses, in order to accomplish and maintain access to these resources. This argument is underlined by Backhouse and Lehmann, who contend that worldwide—but particularly in peripheral regions—new land control frontiers emerge around the expanding production of renewable energies.[16]

Frontiers are either sites of new enclosures and territorialization via the allocation of concessions for resource extraction or the mapping of territories that are decisive for changes in land control.[17] The most novel aspect in this regard is "green framing," i.e., pretending to protect the climate in supposedly already degraded or marginal land, which is employed as a legitimation strategy for new land grabs.[18] A study concerning new frontiers requires an analysis of the different strategies used to structure land control, with the "green" framing of resource extraction particularly interwoven with economic and political dimensions.[19]

Articulations as Analytical Elements

Backhouse proposes the concept of articulation, i.e., the analysis of different elements in a specific chain of meanings, originally theorized by Stuart Hall in the field of cultural studies as an analytical tool that enables the examination of "green" framing and thereby explicitly unfolds the ideological dimension of accumulation processes.[20]

Hall takes advantage of the two components making up the term "articulation." On the one hand, in everyday language, it means to "express" something. On the other hand, it describes a way of linking two elements. Accordingly, articulation is a form of conjunction, which establishes a temporary and contingent unity between distinct discursive elements. Thus, an analysis of articulation can be a useful tool in examining ideology as the articulation of different elements in a specific chain of meanings. Actors play an important role in Hall's concept because individuals are not understood as detached from social relations but have the ability to transform ideology by rearticulating the separate elements and thereby producing a different meaning.

The articulation concept is especially suited to analyzing accumulation processes, since a central element thereof is the separation of producers from the producing means, e.g., the peasant from the territory of interest. Hence, it sheds light on how actors disarticulate (separate) and rearticulate elements. These separation processes are not achieved by pure market forces but by, for instance, specific state programs, violent actions, or legislative changes and are accompanied by a legitimating narrative. The articulation concept helps examine the legitimating narratives that are employed to achieve a lack of contestation.

Previous research has already analyzed discourse strategies and narratives employed by proponents and opponents of large-scale projects, including renewable energy and "green" mining projects.[21] With the aim of promoting land investments, companies frequently articulate their activities with "(sustainable) development and technological progress, modernization, poverty reduction, public welfare, the creation of jobs, and an increase in affluence and income."[22] In the case of "green" mining, scholarship has additionally found a novel association between mining and high-tech industries, "green jobs," and "climate-friendly" extraction. Accordingly, resource exploitation is presented not only as being compatible with climate change but also as necessary to its mitigation.[23] This paper links to this notion by adding empirical insights to the sample from Mexico, which is not only characterized by its contextual particularities but especially highlights the ambivalence of ecological transformation.

The Mexican Mining Sector and Case Selection

The discourse strategies of the most relevant actors in the mining sector in Mexico are presented herein based on an analysis of the largest mining companies and the availability of Global Reporting Initiative (GRI) standards for sustainability (or climate) reports. The Mexican mining sector is divided into some very significant national companies, of which Peñoles—including its subsidiary Fresnillo Plc—is the largest silver producer in the country, while Grupo México holds a quasi-monopoly position in copper production.[24] On the side of foreign companies, the US-based company Newmont, thanks to the immense Peñasquito mining complex, is one of two companies dominating the national extraction of gold, silver, zinc, and lead production. Another firm ranking high in terms of sales is First Majestic Silver, headquartered in Vancouver, representing the disproportionately high share of Canadian companies involved in the Mexican mining market.[25]

The companies' sustainability reports are analyzed in relation to how they aim to highlight their conformity with sustainable development. Additionally, in order to contextualize this with the companies' overall communication strategies, which are the product of professionals who have analyzed in advance the opportunities and risks behind company messaging, the headlines on their homepages are briefly depicted.

The empirical analysis ties up to articulation theory, which provides a methodological framework that avoids monocausal explanations and essentializations.[26] Furthermore, power relations structure possible conditions in the formation of social reality. Consequently, dominance relations always exist in articulated representations, while other possible articulations are suppressed. The following orienting questions relating to articulation analysis led the empirical investigation: How are the elements articulated? What is thematized, and what is being left out?

A key challenge of this form of analysis is its interpretive nature, i.e., structures exist only latently and depend to a large extent on the interpretation of researchers, which is why special care must be taken to meet the requirements of reliability, reproducibility, and validity.[27] According to Mayring's content analysis, these requirements can be met with the help of quantitative and qualitative methods. For this reason, I carried out a quantitative analysis (see Table 10.1 in the Appendix) of the most frequent and most relevant mentions of articulations.[28] Additionally, the qualitative analysis helps to see material not in isolation but as part of a communication chain and thus helps to interpret it meaningfully, based on the theoretical assumptions previously presented.

The selected documents were analyzed through three-step coding via the program MAXQDA. The first step consisted of open coding, namely, a first analytical run through the text when inductively reoccurring motives are coded. The second step is called axial coding. Here, the codes considered relevant were aligned systematically with all the text. By the third step, selective coding, I was occupied with finding core categories through which the codes could be sorted. In addition, during coding, I used the keyword search function in MAXQDA to identify how many times motives reappeared. I excluded keywords in footnotes, endnotes, literature reviews, and appendices, as well as keywords constantly repeated in tables and figures. It is important to mention the reports' different lengths (Peñoles 119, Grupo México 374, FMS 84, and Newmont 62 pages), which led to a deviation in the occurrence of the different codes and categories.

Articulations from Mexican Mining Companies

In this empirical analysis section, I briefly introduce the mining companies by referring to how they describe themselves on their websites, which function as the identity-building gateway of these companies, before turning to the most striking articulations in the analytical codes on *climate change*, *sustainability*, *transition*, *conflict*, *narratives on future*, and *energy provision*.[29] While the results of the quantitative analysis are illustrated in Table 10.1, in the subsections on the companies' articulation strategies, I point to remarkable text passages and noteworthy emphases in the reports.

Peñoles

Peñoles is one of the world's leading silver producers. Its homepage is designed with three banners, each emphasizing the following three pillars of their self-representation. The first headline highlights the long tradition of more than a century as a Mexican company with a global reach in the mining-metallurgical business. The second banner highlights sustainability as a fundamental aspect of its operations. The third headline relates to its code of conduct, which assures integrity and a positive impact on the environment.[30] The company's Sustainable Development Report 2021, written by the company and entitled "Evolving towards a sustainable future," is examined.[31] Peñoles forms part of Grupo BAL[32] and has four main divisions, consisting of explorations, mines, metals, and chemicals.[33] With respect to metals, Peñoles (including its subsidiary Fresnillo plc.) is a leading global producer of refined silver and zinc and a primary actor in lead and gold production in Latin America. The company's most important markets are the United States, Mexico, and Canada.

Within the category *climate change* and its subcodes, the Peñoles report engages comparatively minimally. Its major emphases relate to the *sustainable development* category and focus on different topics, ranging from the compliance and promotion of the UN Sustainable Development Goals (SDGs) over the sustainable development of communities closely located to mining projects, to overall sustainability strategies and self-imposed policies. The categories *transition*, *conflicts*, and *perception of the future* are again under-represented, but the code *RE energy* is another focus of the report.

In the introductory note, the CEO sums up the central message of the report:

> We are committed to responding to the challenge of climate change. Therefore, our emissions mitigation strategy is based on the use of electricity from

> renewable sources, which reduces our emissions and contribute to providing essential metals in the transition to a low-carbon economy.[34]

Throughout the text, it becomes obvious that the report is rather oriented toward compliance in the areas of sustainability and ethical practices; for example, six institutions that certify the company with certain standards are mentioned. All of this aims at supporting Peñoles' claim of being a leader in environmental, social, and governance (ESG) standards, which shows that ESG is considered not a nice-to-have value but a matter of competitiveness that flows into the business strategy.

Grupo México

Grupo México, on its website's home page, presents itself as a "global company with a solid 86-year history."[35] As the largest mining company in Mexico, it generates more than 120,000 direct and indirect jobs, and operations are described as safe and respectful to the environment. The sustainable development report by Grupo México is entitled "We invest in a sustainable future." This information alone offers a narrative that articulates a solid past coupled with a reliable, sustainable future. Grupo México is a Mexico-registered company with a presence not only in the country but also in the United States, Peru, and Spain. Besides its position as a world-leading copper-extracting and producing company, its transportation division offers railroad freight transport in Mexico and the United States. Additionally, the infrastructure division, consisting of operating services in engineering, construction, power generation, exploitation, onshore and offshore drilling, and oil platform leasing, is a significant pillar of the companies' business spectrum, making it the fourth largest company in Mexico in terms of market capital.

Grupo México repeatedly stresses in its report its concerns regarding *climate change-related risks and impacts* and, to a lesser extent, highlights the mitigation actions it implements in its operations. For instance, its use of renewable energy is repeatedly highlighted, with the aim of increasing this through wind energy projects owned by the company.

The report highlights Grupo México's critical role in low-carbon economies due to its leading position in copper extraction:

> In the Mining Division, we mainly produce copper, which, due to its unsurpassed thermal and electrical conductivity qualities, is an essential resource for the manufacture of clean and efficient technologies and allows valuable savings in energy consumption. Copper is vital to low-carbon economies.[36]

The report even mentions the *business opportunity climate change provides*: "Grupo México is one of the world's largest copper producers, so the implications of climate change could benefit the Company's reputation as stakeholders increasingly recognize the importance of copper in helping societies migrate to low-carbon economies."

When searching for the terms "*sustainable/sustainability*" more than 200 results appear in the report, mostly relating to the UN SDGs and the overall goals and policies to move toward sustainable development. In relation to the company's approach to sustainable development, what especially stands out is the effort made in the report to articulate the company's investments in its operations to help the sustainable development of neighboring communities, which is stated over and over again throughout the document. An essential motive in the report is growth, which is again repeatedly articulated alongside community development and is also directly linked to sustainable development and respect for the environment. It is worth noting that "sustainable" or "green mining" is not mentioned anywhere. However, the report shows that the company is aware of its role in the *transition to a low-carbon economy* as a provider of products, especially copper, and services for a "green economy."

What is remarkable about the title of the report, namely, "We invest in a sustainable future," is the positive message about the future that is articulated through sustainability, "green" minerals, and "clean," energy, while in the report the keyword "future" is mostly associated with climate change risks such as longer droughts, which increase water stress and therefore affect the availability of water for operations, which is quantitatively far more relevant but is somehow outdone by the title of the report. Additionally, its compliance with the Paris Agreement and the UN SDGs is outlined by adopting *climate action.*

Sustainable development and community development associated with investments and growth by the company represent another reoccurring articulation. They are presented as factors in a win-win situation. Indirectly, they are linked to the articulation of mining operations and conflicts with affected communities, because overstating the positive impacts on and development of communities can be interpreted as a response to the very present criticism by civil society actors, who claim that their projects cause socio-environmental conflicts.

First Majestic Silver

The headline on First Majestic Silver's (FMS) website reads "There's no substitute for silver,"[37] thereby underscoring its irreplaceable role in industrial applications. FMS is a Canada-registered company and specializes in underground gold and silver mining. It is considered among the principal 14 companies holding mining concessions in Mexico[38] and currently operates seven mines in the country.[39] First Majestic Silver commits itself to comply with the "Towards sustainable mining" commitment made by the Mining Association Canada. The ESG Report is entitled: "The Power of Sustainability."

Climate change is considered in the report as a major threat, while actions directed at mitigating climate change are also stressed repeatedly. At the same time, as in the case of Grupo México, on two occasions climate change is also presented as an opportunity for increased demand in silver for the low-carbon economy. The search for the keyword *sustainable* generates over 50 results. The articulation of *sustainable mining/green mining* is made with the most frequency in comparison with the other reports. One of the main motives in the report is associating the company's investments with "host" community development. With regard to transitioning to a low-carbon economy, the role of silver is repeatedly stressed. Another motive that stands out more compared to the other reports is the positive projection of the future through the company's operations, e.g., by "building a sustainable future."

The headline "There is no substitute for silver" on the website alludes to its function in very relevant applications, such as its use in 5G networks and—most prominently—solar panels. Therefore, it is promoted as a strategic and versatile metal with an increasing global demand due to the worldwide adoption of "green technologies." In the first instance, this is a message to investors, signaling the future-proof business model. The report argues in the same direction by explicitly addressing the "transition to a greener, net zero economy" as a global imperative. For FMS, this implies the redefinition of the mining industry due to the growing demand for critical minerals, as the "low-carbon future will be mineral intensive." In the case of FMS, it is highlighted that "silver [...] is an important material in solar energy applications." Then again, the report also points to the fact that "the mining industry that extracts those minerals must tackle its own carbon-heavy footprint [as] mining accounts for almost two percent of global energy use." Other than, for example, Grupo México and Peñoles, who place most emphasis on the change to renewables as the main source of energy for their operations, FMS only mentions this possibility on one single page. This could be explained by the fact that the former two companies operate their own solar and wind energy divisions, while FMS is a pure mining company. Thus, switching to a renewable energy provision, as limited as this option concerning mining may be, does not include an additional business opportunity for FMS.

FMS enumerates the diverse stakeholders affected by its activities. These range from employees, contractors, unions, suppliers, governments, local communities, indigenous peoples, and advocacy groups through shareholders and investors. FMS's activities are presented as being exclusively beneficial and empowering for stakeholders, with the relationship with "host communities" especially presented as being characterized by collaboration and fair negotiations when it comes to decision-making on matters of land access. In fact, First Majestic Silver has been involved in a publicly reported conflict with the internationally well-known Wixárika people, who conserve their spiritual identity and defend a territory that was incorporated into the sacred natural sites in 1988 by UNESCO.[40] A whole chapter of the report is dedicated to the "responsible government" of their operations. When FMS claims that decisions on land access are based on negotiation and collaboration, they ignore (and thereby conceal) the fact that territorial control is constituted through the firm's governance. This in turn reveals obvious power relations based on capital and other resources that determine the conditions on site.[41]

Although the main theme of the report for long stretches is based on a number of "win-win" scenarios, as a short case study on investment into a community project exemplifies, conflicts with "host communities" are not entirely neglected. It is telling that the main argument for avoiding conflicts is the business risk to the company in terms of damage to reputation, insurance, and legal fees.

Newmont Goldcorp

Newmont chose sliding banners for its website. They all refer to the company's approach to addressing climate change and its responsibility to ESG reporting.[42] Newmont is a US-registered mining company with an emphasis on gold extraction, with by-products also including silver, copper, lead, and zinc. The company has mines in Australia, Africa, and South and North America, among which is Peñasquito mine, which is the largest silver mine in Mexico.[43]

The publishing of a climate report highlights the Newmont communication division's awareness of how important it is to highlight the role their extracted minerals can play in renewable energy technologies. Quantitatively, the concerns on *climate change-related risks* and the company's *actions for climate change mitigation* are most significant in the report. A switch of focus is presented in which "investing in and examining future technical solutions to help address climate change" is proclaimed as the way to go.

The *transition to a low-carbon economy* in the report indeed is a central statement. The first chapter detailing "our role in the transition to a low-carbon economy" points to the importance of produced materials in this transition: "Newmont, and the materials we produce, will play an important role in transitioning the world to a low-carbon economy."[44] Within the report, three transition scenarios in relation to the company's business are presented, each aimed at conveying the impression that the company is fundamentally concerned with climate change mitigation and the energy transition. It is remarkable that when reflecting on "clean" energy transition, the minerals the company produces are cited as copper, silver, and zinc. However, these minerals are considered "co-products," while gold actually generates Newmont's main revenues. In terms of gold sales, the company has seen continued yearly growth over the last three years, and in 2021 gold made up 86% of its overall sales. Fairly enough, in the next paragraph, the role of gold in the transition to a low-carbon economy is mentioned, based on a four-year-old website article by the World Gold Council, a lobby network for the gold industry, directed at investors.[45] This leaves the impression that the report tries to overplay the central focus the company has on gold extraction, which, despite efforts to present it otherwise, has an ecologically extremely damaging footprint[46] and has so far not been considered seriously as a "green mineral."

Another chapter of the report is entitled "Transition Opportunities," in which the company regards itself as sitting in pole position in the company's technological advancements underlining its efforts toward decarbonization in comparison with its market contenders.

Strategically Articulating Mining with the Transition to a Low-Carbon Economy

There is a strong consensus among the analyzed companies in terms of what they understand by contributing to the transition to a low-carbon economy by providing some of the necessary minerals for the expansion of so-called "clean" energy technologies. A difference between the Mexican and foreign companies can be observed in the frequency and clarity with which the argument is articulated. FMS's main message on its website is the irreplaceability of silver in many applications relevant for the electrification of the economy. Newmont, more often than the other companies, stresses the business opportunities the low-carbon transition as a response to climate change offers for the company and its shareholders. Overall, it can be said that in relation to the length of the reports, FMS and Newmont pay more attention than the Mexican companies to the notion that a low-carbon economy depends on the minerals they extract, although the latter organizations are aware of it as well.

In turn, the Mexican companies especially claim to put increased efforts into switching from fossil fuels to "clean" energy in order to run their mining operations.[47] Throughout all the reports, only on one occasion does Peñoles mention that some mining operations processes[48] do not work exclusively with electricity generated by renewable energy.[49] Thus, this is clearly a suppressed matter on which the mining companies prefer to remain silent while they rely on the argument of allegedly "clean" mining as being sufficient to maintain their leading positions in the sector. Additionally, these efforts to increase the use of renewable energy in their own operations are articulated in a narrative in which, due to climate change-related risks, climate change mitigation actions such as switching from fossil fuels to renewable energy must be taken. In this regard, the expansion of "clean" electricity plants such as wind and solar parks, or the growth of the electric vehicle market, is presented in itself as the promotion of decarbonization. This leaves aside the fact that minerals like silver and gold, the extraction of which consumes staggering amounts of fossil fuels,[50] are used only to a very limited degree for renewable energy and energy storage technologies.[51] Additionally, the expansion of "clean" electricity plants does not necessarily correlate with a cutback in fossil fuels, as implicitly argued. As has been examined for Mexico from 2005 to 2019, fossil fuel-based generation has increased three times more than "clean" energy-based generation in the same period.[52]

Another difference around the same distinction line of national and foreign companies is the claim "just transition," which appears in the reports of First Majestic Silver and Newmont. In this regard, Newmont's responsibility in terms of contributing to a *just transition*, vis-à-vis the significant social challenges for vulnerable populations and employment dilemmas due to innovations against the backdrop of transitioning to a low-carbon economy, is admitted. "Just transition" as a claim stems from the analysis that "interrelated systems of capitalism, racism, colonialism, and patriarchy can (re)produce injustices in a less carbon- and resource-intensive economy."[53] Initially, mainly used by trade unions in the Global North, the concept of "just transition" has transcended its origins related to the transformation of labor and the compensation of workers in extractive industries affected by climate policies. As a discourse, "just transition" is the subject of academic debates around environmental, energy, and climate justice.[54] Furthermore, increasingly, it is taken up by civil society organizations and even governmental policymakers, not least in Latin American countries. From the point of view of trade unions, workers, academics, and civil society organizations, it is a problematic development when companies like Newmont adopt terminology that was developed in the first place to raise awareness of the negative effects within a transition, implemented at a cost to the environment and the marginalized communities living in it. The adoption of this terminology demonstrates that in the respective reports the companies seem to be responding to the most recent trends in academic and civil society debates concerning their operations. Moreover, by appropriating such terminology, contestation is made much more difficult because the companies rhetorically pretend to be on the same side as their opponents. Thereby, not only are the socio-environmental costs of mining operations made invisible, but they are also normalized by the assumption that they are necessary to combat climate change.

Promoting Investments with Sustainability and Community Development

The empirical analysis reveals that companies mostly maintain their focus within their reports on sustainability, which may be related to the development of the Global Reporting Initiative (GRI), set in motion around the turn of the century as a response to the 1987 Brundtland Report. In this regard, companies were addressed as furthering "sustainable development," defined as "development that meets the needs of the present without compromising the ability of future generations to meet their own needs."[55] Ever since, many companies of the function and size presented herein have committed themselves to publishing GRI-certified sustainability reports. Whether this practice has led to a higher degree of accountability or even compliance with sustainable development as defined by the Brundtland Report has been controversially debated for roughly two decades now.[56]

In contrast to the research mentioned before, in which company narratives were analyzed, articulations of "green jobs" or the "high-tech industry" were hardly present in the reports, thereby demonstrating the need for a context-specific analysis. Otherwise, a common line with previous research is the emphasis placed by the analyzed companies (sustainable) on the development, progression, and welfare of neighboring communities. Closely related to this notion is the articulation of investment and growth with sustainable development, while community support is quantitatively more prominent in comparison to energy transition. This argument positions mining investments as delivering riches, growth, and job opportunities. Thus, it draws on traditional economic discourses articulating a nexus of mining with development, wealth, and social progress for all "stakeholders."[57] This link, however, lacks any empirical evidence, as research on socio-economic development in municipalities surrounding mining projects in Mexico shows that economic living conditions there have not improved. On the contrary, in 13 municipalities the poverty rate is below the national average, and in seven municipalities a major level of extreme poverty prevails.[58]

By stressing collaborative relationships with "host communities" and the aim of increasing value for all stakeholders,[59] i.e., a motive all reports except for Newmont[60] adopt, the reports address shareholders expecting profit distribution and a "clean" image for their investments. On a meta-level, it can be interpreted that these messages are also directed at criticisms made by environmentalists, accusing the companies of being responsible for socio-environmental conflicts, and furthermore at academics that are charging the sector for the little socio-economic benefit neighboring communities see within this enclave-economies model adopted by most mining projects.[61]

The companies' emphasis on investing in the sustainable development of neighboring communities, which is mentioned especially by Grupo México and First Majestic Silver, can be interpreted as a way of obscuring the negative environmental and social impacts large-scale mining has on its surroundings. It is no coincidence that the category *conflicts with communities*, although briefly mentioned in all reports, is more frequently thematized by the same companies. Both have been prominently involved in publicly famous mining conflicts. In 2015, for instance, Grupo México's main mine Cananea released highly toxic substances that ran into the Sonora River, affecting around 22,000 persons.[62] Moreover, as already mentioned, First Majestic Silver was involved in a years-long conflict with the indigenous group Wixárika.

Conclusion

By applying the concept of articulation analysis to sustainability reports published by the most relevant mining companies in Mexico, I have examined the discourse strategies employed to promote acceptance of mining. As revealed herein, these discourse strategies used by the mining companies very explicitly address the subjects of climate change, sustainable development, transition to a low-carbon economy, etc. The articulations found therein suggest an increase in ecological awareness in Mexican civil society in relation to mining conflicts, the environmental and social costs of mining projects, and the urgency of ecological transformation.

The discourse strategies are applied within an ongoing framework of mining expansion in which the pertinent threats of climate change are articulated as economic growth opportunities. Rather than transitioning within the sector toward extracting so-called "green" minerals, the articulation analysis indicates that the discursive adoption of the energy transition is integrated into the already entrenched heuristics of (sustainable) development, technological progress, and economic growth. Therefore, the adoption of these rhetorical articulations can be interpreted as a further way in which mining corporations promote acceptance of their operations, in order to continue making profit from mining operations and to maintain their powerful market positions. This is exemplified by the fact that in terms of the value generated via its extraction and production, gold remains the most important mined mineral in Mexico, though it is not a significant metal for renewable energy technology.

Coming back to the question concerning existing ambivalences toward ecological transformation, the articulation analysis of the main mining companies in Mexico proves their awareness of their role as providers of minerals for the transition to a low-carbon economy. However, this role varies from ambivalent, due to the aforementioned awareness, to absolutely destructive in light of an urgently needed ecological transformation. This is the case because the promoted model of transition is one based on a massive expansion of the mining frontier, which by no means entails decarbonization. The fact that some processes within mining extraction and production are not feasible with renewable energy is equally suppressed within the reports as the requirements of fossil fuels for solar and wind energy infrastructure. A reduction in greenhouse gases, by shifting toward renewable sources for mining operations, is proposed by the companies in a way proportional to the imperative of an overall increase in production, which, of course, in sum does not reduce but actually elevates greenhouse gas emissions. This means that by rhetorically promoting a transition to a low-carbon economy in Mexico—and certainly not only there—in fact, a proportional increase in renewable energy within an ever-expanding fossil energy matrix is promoted.

This contribution by no means aims at discrediting the transition to decarbonization, in which renewable technologies definitely play a significant role. However, the fact that the same actors responsible for serious environmental damage now sense the opportunity to "save the world" from climate change catastrophe with new technologies should be cause for caution. The analysis undertaken herein illustrates the ambivalence, or rather contradictions, within the ecological transformation where companies proclaim technical change without addressing politics. Instead, I argue that the present analysis should be seen as part of a larger debate on these technologies as part of an extraction process under the logic of dispossession, unequal power relations, and the concentration of profits. In other words, by explicitly addressing the political core of such technological changes, I am speaking to debates led in political ecology, political economy, or ecological economics on the social and environmental costs of ecological transformation. The analyzed discourse strategies present an important element in the ongoing processes of dispossession, concentration of profits, and expansion of the extractive frontier. A "just transition" implies not only the technology used but also the transformation of overcoming the logic to satisfy the profits of a few—i.e., large mining companies, and instead it is subject to questions related to how energy distribution for the many is negotiated without destroying their livelihoods and those of future generations.[63]

Notes

1 Daniele La Porta Arrobas et al., *The Growing Role of Minerals and Metals for a Low Carbon Future* (Washington, DC: The World Bank Group, 2017), https://openknowledge.worldbank.org/bitstream/handle/10986/28312/117581-WP-P159838-PUBLIC-ClimateSmartMiningJuly.pdf?sequence=1&isAllowed=y.

2 Meadhbh Bolger et al., *'Green mining' is a myth: The case for cutting EU resource consumption* (Brussels: European Environmental Bureau, 2021), https://friendsoftheearth.eu/wp-content/uploads/2021/10/Green-mining-myth-report.pdf.

3 Susana I. Velázquez Quesada et al., *Explotación de litio en México: ¿Interés público o extractivismo transnacional?* (Mexico-City: REMA, 2023), www.remamx.org/wp-content/uploads/2023/06/Informe_Litio_REMA_MWC_20233.pdf.

4 CAMIMEX, *Informe Anual Camimex 2022* (Mexico-City: CAMIMEX, 2023), https://camimex.org.mx/application/files/8816/6247/9890/info2022.pdf.

5 Kirsten Hund et al., *Minerals for Climate Action: The Mineral Intensity of the Clean Energy Transition* (Washington, DC: World Bank Group, 2020), https://mineral-choices.com/content/minerals-for-climate-action/.

6 CRU International, *The Role of Silver in the Green Revolution: Prepared for The Silver Institute* (Cranberry Woods: Silver Institute, 2018), www.silverinstitute.org/wp-content/uploads/2018/07/Role_of_Silver_Green_Revolution_28Jun2018.pdf.

7 Aleida Azamar Alonso, "El Litio en México: Verdades y Mentiras," in *Minería en México: panorama social, ambiental y económico,* ed. Aleida Azamar Alonso and Isidro Téllez Ramírez (Mexico-City: SEMARNAT, 2022), 27–45.

8 "Financiamiento para organizar Plataforma para el Diálogo," *CALAS*, last modified February 14, 2023, http://calas.lat/convocatorias/financiamiento-para-organizar-plataforma-para-el-di%C3%A1logo.

9 James Fairhead, Melissa Leach, and Ian Scoones, "Green Grabbing: a new appropriation of nature?" *Journal of Peasant Studies* 39, no. 2 (2012): 237–261.

10 Maria Backhouse and Rosa Lehmann, "'New 'renewable' frontiers: contested palm oil plantations and wind energy projects in Brazil and Mexico'," *Journal of Land Use Science* 15, no. 2-3 (2020): 373–388.

11 Nancy L. Peluso and Christian Lund, "New frontiers of land control: Introduction," *Journal of Peasant Studies* 38, no. 4 (2011): 667–681.

12 Daniel M. Voskoboynik and Diego Andreucci, "'Greening' extractivism: Environmental discourses and resource governance in the 'Lithium Triangle'" *Environment and Planning E: Nature and Space* 5, no. 2 (2021): 787–809.

13 Christoph Görg, *Regulation der Naturverhältnisse: Zu einer kritischen Theorie der ökologischen Krise* (Münster: Westfälisches Dampfboot, 2003).

14 Gavin Bridge and Phil McManus, "Sticks and Stones: Environmental Narratives and Discursive Regulation in the Forestry and Mining Sectors," *Antipode* 32, no. 1 (2000): 10–47.

15 Voskoboynik and Andreucci, "'Greening Extractivism'."

16 Backhouse and Lehmann, "New 'renewable' frontiers."

17 Peluso and Lund, "New frontiers of land control."

18 Saturino Borras and Jennifer Franco, "Global Land Grabbing and Trajectories of Agrarian Change: A Preliminary Analysis," *Journal of Agrarian Change* 12, no. 1 (2012): 34–59.

19 Backhouse and Lehmann, "New 'renewable' frontiers."

20 Original accumulation, or primitive accumulation, is a process described by Marx in the first volume of "The Capital." It includes "privatization of land and the forceful expulsion of peasant populations; the conversion of various forms of property rights (common, collective, state, etc.) into exclusive private property rights; [...] and the suppression of alternative (indigenous) forms of production and consumption; colonial, neo-colonial, and imperial processes of appropriation of assets (including natural resources)." David Harvey, *The New Imperialism* (Oxford: Oxford University Press, 2003). "Harvey suggests that accumulation processes have been particularly significant during the past 30 years, in the context of International Monetary Fund – and World Bank – led structural adjustments. Indeed, his description of these processes corresponds with the neoliberal restructuring of Mexico's mining sector." Darcy Tetreault, "Social Environmental Mining Conflicts in Mexico," *Latin American Perspectives* 42, no. 5 (2015): 44–66; Maria Backhouse, "Ursprüngliche Akkumulation und Ideologie. Impulse von Stuart Hall," in *In Hörweite von Stuart Hall: Gesellschaftskritik ohne Gewähr,* ed. Maria Backhouse, Stefan Kalmring, and Andreas Nowak (Hamburg: Argument, 2017), 49–66.

21 Kristina Dietz and Bettina Engels, "Field of Conflict: Ein relationaler Ansatz zur Analyse von Konflikten um Land," *(GLOCON) Global Change – Local Conflicts Working Paper* (2018).
22 Kristina Dietz and Bettina Engels, "Analysing land conflicts in times of global crises," *Geoforum* 111 (2020): 208–217.
23 Voskoboynik and Andreucci, "'Greening Extractivism'."
24 Azamar Alonso, "El Litio en México."
25 Isidro Téllez Ramírez and Maria T. Sánchez Salazar, "Concesiones mineras y acaparamiento del subsuelo," in *La dimensión global de las regiones y sus reconfiguraciones económicas y urbanas*, ed. P. Wong González et al. (Mexico-City: UNAM, 2021).
26 Marc Drobot, "Stuart Halls ›Theorie der Artikulation‹: Eine Rahmenmethodologie für die Protest- und Bewegungsforschung," in *Handbuch Poststrukturalistische Perspektiven auf soziale Bewegungen*, ed. Judith Vey, Johanna Leinius and Ingmar Hagemann (Bielefeld: transcript Verlag, 2019), 131–248.
27 Michael Oswald, "Die Methode der Frame-Analyse," in *Strategisches Framing: Eine Einführung*, ed. Michael Oswald (Wiesbaden: Springer VS, 2019), 169–75.
28 Philipp Mayring and Thomas Fenzl, "Qualitative Inhaltsanalyse," in *Handbuch Methoden der Empirischen Sozialforschung*, ed. Nina Baur and Jörg Blasius (Wiesbaden: Vieweg, 2019), 633–48.
29 Subsequently, codes and subcodes are written in italics.
30 "Industrias Peñoles," Home, Industrias Peñoles, last modified February 18, 2023, www.penoles.com.mx/en.
31 Industrias Peñoles, *2021 Sustainable Development Report* (Mexico-City: Industrias Peñoles, 2022), www.penoles.com.mx/assets/files/reportes/Anuales/Desarrollo%20Sustentable/en/PEN_IDS21-ENG.pdf.
32 BAL stands for a cluster of companies engaging in very different forms of businesses, such as pension funds (Profuturo), insurance (Grupo Nacional Provincial), mining (Peñoles, Fresnillo), metallurgical and chemical industries (Peñoles), financial services (Valmex Soluciones Financieras), agricultural businesses (Agropecuaria Internacional), education (Instituto Tecnológico Autónomo de México), exploration and hydrocarbons production (PetroBal), and power generation (Energía Eléctrica BAL). Industrias Peñoles, "*2021 Sustainable Development Report.*"
33 Exploration (of deposits in Mexico and South America), mines (four underground mines producing zinc, lead, and copper), metals (metallurgical operations that process metals from company-owned mines, subsidiaries, and other mining companies, producing refined metals in Coahuila and Durango (Met-Mex in Torreón), and chemicals (the main products are sodium sulfate, magnesium oxide, and magnesium sulfate). Grupo México, *We invest in a sustainable future: Sustainable Development Report 2021* (Mexico-City: Grupo México, 2022), 245, www.gmexico.com/GMDocs/InformeSustentable/Eng/IDS2021_ING.pdf.
34 Industrias Peñoles, "2021 *Sustainable Development Report*," 5.
35 "We are a global company with a solid 86-year history," *Grupo México*, last modified February 18, 2023, www.gmexico.com/Pages/default.aspx.
36 Grupo México, *We invest in a sustainable future*, 245.
37 "First Majestic Silver—There's no substitute for silver—Home," *First Majestic Silver*, last modified February 18, 2023, https://firstmajestic.com/.
38 Téllez Ramírez and Sánchez Salazar, "Concesiones Mineras."

39 First Majestic Silver, *The Power of Sustainability: 2021 Sustainability Report* (Vancouver: First Majestic Silver, 2022), https://firstmajestic.com/_resources/reports/FM-2021-ESG.pdf?v=0.269.
40 "Conflicto Minero: Pueblo Wixárika defiende su territorio sagrado Wirikuta de la destrucción minera," *OCMAL*, last modified February 24, 2023, https://mapa.conflictosmineros.net/ocmal_db-v2/conflicto/view/202.
41 Peluso and Lund, "New frontiers of land control."
42 "Newmont Corporation—Home," *Newmont*, last modified February 28, 2023, https://newmont.com/.
43 Newmont, *Leading Change. Focused on Value: 2021 Climate Report* (Colorado: Newmont, 2022), https://s24.q4cdn.com/382246808/files/doc_downloads/sustainability/2021-report/2021-Climate-Report.pdf.
44 Newmont, *2021 Climate Report,* 6.
45 "World Gold Council. Gold's role in a low-carbon economy," *Gold*, last modified February 16, 2023, www.gold.org/goldhub/research/gold-investor/gold-investor-february-2019/13644.
46 "To obtain 1 ounce of gold requires the removal of 150 tons of rock, 40 kilograms of explosives, consumes 150 to 200 thousand liters of fresh water (enough to provide water for a family per year), 500 liters of fuel, emits 650 kilograms of CO_2 into the atmosphere, along with other highly polluting and toxic greenhouse gases" (own translation; Rodolfo García Zamora and Selene Gaspar Olvera, "El extractivismo minero en México bajo la Cuarta Transformación 2018–2024," *Revista nuestrAmérica* 8, no. 16 (2021)).
47 It is worth mentioning in this regard that in Mexico, aside from solar and wind energy, nuclear energy and other energy generation forms are also considered "clean." Yannick Deniau et al., *Alumbrar las contradicciones del Sistema Eléctrico Mexicano y de la transición energética: Preguntas claves para entenderlos y construir otros modelos energéticos* (Mexico-City: GeoComunes, 2021), http://geocomunes.org/Analisis_PDF/GC_SistemaElectrico_2020.pdf.
48 For example, in the case of lithium calcination. Temperatures above 950 Celsius require gas burners. Velázquez Quesada et al., *Explotación de litio en México.*
49 Likewise, solar and wind infrastructure requires the consumption of minerals and hydrocarbons in manufacturing generators or panels, building a plant, maintaining it, and decommissioning the plant after its lifetime. *Deniau et al., Alumbrar las contradicciones.*
50 See endnote 46.
51 For example, in 2015, solar PV accounted for nearly 7% of total silver demand (Hund et al., *Minerals for Climate Action.*
52 Deniau et al., *Alumbrar las contradicciones.*
53 Erik Kojola and Julian Agyeman, "Just Transitions and Labor," in *Handbook of Environmental Sociology*, ed. Beth Schaefer Caniglia et al. (Cham: Springer, 2021), 115–138.
54 Kirsten E. H. Jenkins et al., "Energy justice: A conceptual review," *Energy Research & Social Science* 11 (2016): 174–182.
55 José M. Moneva, Pablo Archel, and Carmen Correa, "GRI and the camouflaging of corporate unsustainability," *Accounting Forum* 30, no. 2 (2006): 121–137.
56 Marcio C. Machado et al., "Can Global Reporting Initiative reports reveal companies' green supply chain management practices?" *Journal of Cleaner Production* 383 (2023).
57 Voskoboynik and Andreucci, "Greening Extractivism."
58 Azamar Alonso, "El Litio en México."
59 First Majestic Silver, *2021 Sustainability Report.*

60 This can be explained by the fact that the climate report was analyzed instead of the sustainability report.
61 Azamar Alonso, "El Litio en México."
62 "La Cananea del río Sonora," *OCMAL*, last modified February 24, 2023, www.ocmal.org/la-cananea-del-rio-sonora/.
63 Deniau et al., *Alumbrar las contradicciones*; Zamora and Olvera, "El extractivismo minero."

Appendix

Table 10.1 Quantitative keyword analysis

	Peñoles	*Grupo México*	*Newmont*	*First Majestic Silver*
• Climate Change (CC)	7	46	68	25
CC-related risks and impacts	3	31	27	12
CC (mitigation) action	4	14	35	11
CC as a business opportunity	/	1	6	2
• Sustainability	65	47	8	25
Sustainable/green mining	3	/	2	5
Sustainable development	59	17	6	1
Sustainable and community development and investment/growth	3	30	/	19
• Transition	3	14	15	14
Transition to a low-carbon economy	2	10	8	6
Just transition	/	/	7	1
Transition/green minerals	1	4	/	7
• Conflicts	2	3	1	4
Conflicts with neighboring communities	2	3	1	4
• Perception of the future	2	3	3	10
Green/clean/sustainable future	2	3	3	10
• RE Energy	21	25	8	12
Renewable energy as a source for operations	13	19	/	3
"Clean" energy	8	6	8	9

Source: Copyright Rafael Hernández Westpfahl.

11 Integrating Labor, Environment, and Climate?

(Dis)connections in the Spanish and Portuguese Energy Decarbonizations

Rocío Hiraldo

Introduction

The decarbonization of the Spanish and Portuguese energy sectors is contributing significantly to attaining the ambitious climate change mitigation targets set in 2020 by the European Green Deal.[1] Between 2018 and 2022, 12 coal-fired power plants were shut down in Spain[2] and 2 in Portugal, together with an oil refinery.[3] Moreover, the renewable energy sector has been growing rapidly in both countries. In 2022, more than 58% of the total energy generated in Spain came from renewable energy sources, while in Portugal, 78.0% was recorded in December 2023.[4]

Despite their contribution to climate change mitigation, these developments are strongly criticized by labor, environmental, climate, and other social movements. These mobilizations have the potential to push government and private companies to implement energy transitions that are considerate of climate, environmental, and labor issues. However, previous research indicates tensions between the objectives these movements pursue.[5] Due to the prevalence of economic models highly reliant upon environmental and climate-deleterious industries, the defense of labor conditions by workers and trade unions has often come at the expense of environmental protection. Similarly, environmental non-governmental organizations (NGOs) have often been oblivious to the labor implications of environmental protection and climate change mitigation measures.[6] At the same time, the literature has well documented the role of low-income workers (i.e., people who live through their own labor) and labor organizations in struggles against environmental degradation,[7] as well as the various alliances between these actors and environmental organizations.[8] The continuity of this convergence between mobilizations by labour, environmental and climate activists is exemplified by the growing use of the term "Just Transition" (JT) (i.e., the idea that ecological and low-carbon transitions should incorporate labor concerns) by these actors and their involvement in the creation of climate job campaigns in the last decade.[9]

Taking these contradictory processes into consideration, this chapter explores the following questions. First, how are labor, environmental, and climate goals being integrated in mobilizations around the decarbonization of the Spanish and Portuguese energy sectors? Second, is (and, in affirmative case, in what contexts) ambivalence a suitable strategy for compensating tensions around such

DOI: 10.4324/9781032627984-14

integration? The chapter studies these questions by focusing on mobilizations around coal-fired power plant and oil refinery closures, as well as large-scale renewable energy projects developing in these two countries since 2020. To do so, it asks three questions: (a) how and why do workers (i.e., people who live through their own labor, both waged and unwaged) and labor organizations (not) engage with environmental and climate change concerns?; (b) how and why do environmental NGOs (not) integrate labor and climate issues into their demands?; and (c) how and why do climate organizations (not) incorporate labor and environmental concerns in their mobilizations? This study is based upon the collection of secondary data (organizations websites, reports, and news) aimed at mapping activist organizations' demands and actions in the contexts studied. In the case of Spain, data collection focused on the Andalusia and Galicia regions, where energy sector decarbonization has rapidly developed in the past five years. Additionally, primary data were collected through ten interviews with unionists, workers, rural dwellers, and environmentalists mobilizing in the context of energy decarbonization in Galicia (Spain).

The next section reviews the literature concerning the convergence of environmental, labor, and climate goals in social mobilizations. This is followed by an analysis of the case studies. The chapter concludes by reflecting on the theoretical and political implications of this research.

Ambivalences Around the Integration of Labor, Environmental, and Climate Goals in Social Mobilizations

The integration of labor, environmental, and climate issues in social mobilizations has often been mapped through the study of the alliances between trade unions, workers, and environmental organizations, also known as "red-green" and "blue-green" coalitions.[10] While the existence and persistence of these partnerships is significant in the analysis of such integration, social scientists and activists have challenged the portrayal of environmental and labor mobilizations as compartmentalized struggles joining forces temporarily and strategically. In so doing, they have demonstrated how environmentalism is not exclusively a concern of (middle-class) environmentalists and environmental NGOs. In the Global North and South, workers, labor organizations, and working-class (often racialized) communities have engaged in various struggles against pollution, ecosystem degradation, and exclusionary conservation. This has led to the coining of terms such as "labor environmentalism," "environmentalism of the poor," "working-class environmentalism," and "people of color environmentalism."[11] Nonetheless, workers and unions' environmentalism is ambivalent. A number of material challenges result in workers' reluctance to support environmental and climate protection measures, among which are job blackmail (i.e., the fear of losing their jobs), inferior wages in green economic sectors in comparison to environmentally polluting ones, and assumptions that environmentally deleterious industries are more prone to creating jobs. There are also cultural challenges such as the protection of men's identities as breadwinners and the association of working-class identities with environmentally destructive sectors such as mining.[12] Similarly, unions also face a number of difficulties when supporting environmental and climate policies, because while in some cases they may support these programs, they may also have opposing and hedging attitudes. This ambivalence is connected to the fact that they are representative organizations and, more specifically, to the strategies they employ to maintain and increase their power resources. Thus, unions are often supportive of environmental protection and climate change mitigation policies that converge with their sectorial interests, while they are opposed to policies negatively impacting on the workers they represent.[13] At the same time, their attitudes toward environmental and climate protection are also connected to their engagement with environmental issues in the past, often shaped by individuals within organizations.[14] Nonetheless, unions' road to labor environmentalism is not linear but rather full of ambivalences and contentions.[15]

Environmental activists and environmental NGOs also have ambivalent attitudes toward labor issues and working-class environmentalism. While some environmental and climate NGOs have incorporated a labor dimension in their discourses, activists within these organizations are not necessarily concerned by, or aware of, working-class people's problems.[16] Moreover, various studies have shown that these organizations have engaged in environmental conservation and climate change mitigation initiatives, with negative effects on the labor and living conditions of peasants.[17] At the same time, throughout history, certain environmental and climate NGOs have shown support for workers and unions' struggles and have engaged in red-green alliances demanding green and climate jobs.[18]

The integration of labor, environmental, and climate goals is also affected by a broader political economy. Policies allowing deindustrialization and capital mobility can reinforce job blackmail, thereby making unions and workers more reluctant to consider environmental regulation. Moreover, neoliberal governments' lack of funding for environmental protection is said to foster environmentalists' prioritization of fundraising over grassroots organization through alliances with workers and unions.[19]

The connection between labor, climate, and environmental mobilizations is rendered even more complex by the diversity of ways in which it is conceived. This has become increasingly evident following the more recent popularization of the JT concept among unions, environmentalists, researchers, and policymakers. Some approaches emphasize private companies' role as providers of jobs and infrastructures in green and low-carbon transitions, whereas others focus on government-led public-sector approaches to JT. Both, however, generally advocate for a shared solution approach focused on social dialog as a strategy for JT. Other approaches see the reversal of power inequalities as central to JTs, thus advocating for grassroots and transformative justice approaches to environmental and climate protection.[20]

Labor-Environmental-Climate (Dis)connections in Spanish and Portuguese Energy Decarbonization

This section analyzes ambivalences in the connection between labor, environmental, and climate goals in mobilizations around the decarbonization of the Spanish and Portuguese energy sectors.

Focusing on Immediate Needs: Workers and Unions Facing Closures

The example of coal-fired power plants and oil refinery closures emphasizes the argument that material challenges are central to workers' and unions' ambivalent support of environmental and climate change policies. The lack of (adequate) employment alternatives after these closures led unions and workers to put forward labor demands, even though they may have been counterproductive for energy decarbonization.

In 2021, the European Union created a Just Transition Fund (EUJTF) aimed at supporting countries in their transition toward climate neutrality.[21] This fund provides the Spanish and Portuguese governments, for instance, with economic support for local areas affected by the closure of coal-fired power plants and oil refineries. Earlier on, in 2019, the Spanish government initiated a JT financing mechanism to facilitate the creation of green energy projects and employment alternatives for workers affected by the closure of coal-fired power plants in the country. However, energy companies closed coal-fired power plants and oil refineries prior to the implementation of the EUJTF. The closure also happened before the implementation of green energy projects aimed at providing economic alternatives for affected areas.[22] More importantly, neither auxiliary workers employed by sub-contracted companies nor those whose petty businesses depended on the functioning of the plant received support after a closure.[23] As Table 11.1 illustrates, in most of the plants and oil refineries studied, these two categories of workers were higher in number than direct employees, meaning most workers did not receive alternatives.

Furthermore, according to interviews conducted for this research, in several cases, direct employees in As Pontes were given the option to either take an alternative position at a plant outside Galicia (the region where As Pontes is located) or leave the company. Affected workers looking after their relatives (children and elderly people) had no option other than to leave the company.

Table 11.1 Approximate numbers of workers affected by the closures

Plants and refineries	*No. of direct employees*	*No. of auxiliary workers*	*No. of induced jobs*	*Total job losses*
Portugal				
Matosinhos	401	1000	5000	6401
Pego	83	90	N/A[a]	173
Sines	109	400	90	597
Spain				
As Pontes	108	327	743	1178
Meirama	224	60	N/A	284
Carboneras	137	269	N/A	406
Los Barrios	110	40	N/A	150
Puente Nuevo	70	150	N/A	220

[a] Numbers that are not available are presented as N/A.

Source: Copyright Rocío Hiraldo.

In this context of a labor emergency being added to the climate emergency, workers and unions organized various mobilizations, such as protests in front of the plants, roadblocks, meetings between unions and energy companies, and sit-ins.[24] They criticized not only the lack of adequate alternatives for all workers, but also the speed of dismissals, the abruptness of the closures, and the lack of anticipation and planning. To compensate for these issues, workers and unions called for extensions to the functioning of coal-fired power plants and oil refineries, and even their reopening[25] despite the fact that before and after closures, unions in these sectors demonstrated a concern for climate change through various actions such as workshops and meetings.[26]

In Portugal, the General Confederation of Portuguese Workers (CGTP in Portuguese) criticized the abrupt decision to shut down the Matosinhos oil refinery and the Pego coal-fired power plant. The union also called for a delay to the closure until "there [was] a credible alternative solution that actually ensure[d] the preservation of all jobs," stressing that "a change of this magnitude… must be preceded by alternatives."[27]

Spanish unions and workers took a similar stance. Following the energy crisis generated by Russia's war against Ukraine, the Galician Inter-union Confederation (CIG in Galician) called for the full reactivation of the coal-fired power plant in As Pontes.[28] Earlier, in September 2020, auxiliary workers at the coal-fired power plant in Los Barrios, supported by General Workers Union (UGT) representatives, demanded a delay to the closure until all the stored coal was burnt, alongside workers' continued involvement in this process.[29] Notwithstanding, in February 2022, when Viesgo (the plant owner) announced the reopening of the coal-fired power plant, the UGT took a different position and discouraged workers from going back to work due to the unclear monetary assurances for them. Moreover, it criticized the company for its opportunism in terms of its relationship with its workers:

> … the company [Viesgo] makes an ERE [a massive layoff] to fire workers because the megawatt per hour (MWh) was at 20 euros, but now that it is at 200 euros, they call workers because they expect a great deal…[30]

As part of their strategy to protect local employment, some workers and unions have defended the implementation of green hydrogen and biomass energy projects criticized by environmental organizations for not being clean energy sources.[31] In a press release, workers at the Meirama power plant, for instance, regretted Naturgy's refusal to convert the facility into a biomass energy plant. In Carboneras, the UGT demanded financial support in order to substitute coal for biomass, and in September 2023, the CIG emphasized the need to reindustrialize As Pontes as a way of compensating labor costs following the closure of the coal-fired power plant in the town. After operations at the coal-fired power plant in Pego were stopped, the CGTP reminded the press about previous plans for a biomass energy project—and its potential in the creation of employment alternatives at the local level.[32]

Engaging Ambivalently with Labor in a Climate Emergency

Environmental and climate NGOs in Spain and Portugal have been ambivalently attentive to the labor emergency generated by the closure of coal-fired power plants and oil refineries in these countries. Such ambivalence can be seen as a strategy influenced by the nature of these organizations and its manifestation in a context of growing popularization of JTs. Unlike in the case of unions, the legitimacy of these organizations does not depend so much on their responsiveness to workers; instead, they respond to individuals and institutional donors who may not necessarily be concerned about the labor implications of energy decarbonization. However, as activist organizations, their survival relies on their adaptation to changes in policy and social discourse, too.[33] As the following paragraphs illustrate, Spanish and Portuguese environmental and climate NGOs have included JT in some of their discourses and demands. Nonetheless, their engagement with the labor implications of coal-fired power plant and oil refinery closures has only been partial, albeit with some variation between organizations.

Central in most recent mobilizations against coal-fired power plants and oil refineries in Spain and Portugal has been the Europe Beyond Coal (EBC) campaign, launched in July 2017. The EBC was led by a network of environmental NGOs, among others the World Wide Fund for Nature (WWF), Greenpeace, Friends of the Earth, and Zero, a Portuguese climate organization. More than 20 protests were organized to demand the closure of coal-fired power plants in Spain by 2025. The campaign also included the screening of a documentary called "The Dark Side of Coal," produced by Greenpeace.[34] This organization was one of the most active in the Spanish EBC campaign, carrying out a number of actions a few weeks before the Conference of Parties (COP) of the United Nations Framework Convention on Climate Change in December 2018. These protests consisted mostly of unfolding banners next to power plants.[35] During these mobilizations and, more broadly, in the context of coal-fired power plant closures, Greenpeace was ambivalent toward the incorporation of labor goals, with labor issues notably absent in messages displayed during the campaign. Moreover, the organization was not present at protests organized by trade unions and affected workers. In addition, in a report on the transition to a coal and nuclear-free energy system published in March 2018, the organization only mentioned labor issues once, and this was only as a suggestion that the closure of nuclear plants could lead to the creation of 300,000 jobs in the renewable energy sector. There was no mention about the need to create adequate employment alternatives.[36] Nonetheless, in several press releases and blog posts, Greenpeace discussed the labor implications of coal-fired power plant closures. More specifically, it demanded a social dialog-based plan for the abandonment of dirty and dangerous energies in order to ensure a just energy transition for workers.[37]

The Galician EBC campaign, called "Galiza, un futuro sen carbon," has also been key in recent environmentalists' struggles against fossil fuels in this region. Neither workers nor trade unions were part of the platform, which was formed by 14 environmental organizations.[38] However, some of the demands of the Galician EBC campaign,[39] which Greenpeace published on its website, included the need to take JT measures for workers—not only those employed at the power plants, but also lorry drivers. This broader approach to labor may be connected to the fact that the platform was formed by local and national environmental NGOs with a less biocentric approach toward conservation and climate change mitigation. Among these were the Association for the Environmental Defense of Galicia (Adega), the most important environmental organization in Galicia, and Ecologistas en Acción (EA), a confederation of more than 300 environmental groups from all over Spain.[40] Greenpeace and these two organizations rejected the delay to coal-fired power plant closures and applauded the final decision to shut down the plants, even when decisions were taken abruptly.[41] At the same time, in line with Greenpeace, these organizations complained about the lack of planning and the need to consider job losses, provide alternatives for affected workers, and create a roadmap to do so. Their demands reflected a deeper incorporation of the labor and, more broadly, social dimensions of coal-fired power plant closures than those of Greenpeace. Adega slated Endesa for forcing workers at As Pontes to pay the costs of the transition, while highlighting this company's historical role in looting the area through the displacement, appropriation, and environmental destruction of the region. EA also denounced the uncertainty surrounding auxiliary workers and the negative consequences of the closure for lorry drivers, whom the organization characterized as "fake self-employed," meaning Endesa benefitted from them but externalized onto them the costs of hiring labor. Ecologistas en Acción-Verdemar (EA's branch in Los Barrios) criticized the central government for its continuous mistreatment of the Andalusian region, namely destroying its local economy, not only through fossil fuel energy schemes, but also renewable ones:

> Campo de Gibraltar [the area where Los Barrios is located] continues to be "the Cinderella" of Andalusia, depending on fossil fuels and "uprooting the local economy."[42]

This organization was present during the EBC campaign in Andalusia, showing a stronger connection to workers than Greenpeace. As the campaign developed in Almería, the local EA delegation expressed its willingness to work together with workers during the closure of the coal-fired power plant in Carboneras.[43] However, no work developed further thereafter, and the organization was not seen at protests attended by workers at the plant.

In Portugal, two environmental organizations have been key in the struggle against coal-fired power plants and oil refineries: Zero and Climáximo. In line with Spanish environmental NGOs, their engagement with labor issues has been ambivalent. Zero, for instance, participated in the EBC campaign and accentuated the need for workers' participation in decision-making processes around closures, as well as the provision of adequate and sustainable employment alternatives for them.[44] Nonetheless, the organization was not present at workers' protests and did not develop alliances with them or their unions' representatives. Similarly, Climáximo, the main Portuguese climate organization, demonstrated clear concerns about the implications of closures for workers, citing particularly the need to integrate them in decision-making related to power plant closures and demanding the elaboration of "a just transition plan based on a social dialog that favors workers and affected communities." Members of the organization carried out JT training sessions for other climate activists at protests in front of coal-fired power plants and oil refineries. Furthermore, they criticized the Portuguese Government for shifting the meaning of JT as a profit-driven process ignoring workers' needs:

> Instead of being used as a set of guarantees for workers' rights in polluting sectors, the term "Just Transition" is being used as a set of guarantees for shareholders, [in order to maintain] their profits.
>
> (Climáximo, 2023)[45]

This inclusion of labor issues is connected to Climáximo's coherence with its political trajectory as the former leader of a Portuguese climate jobs campaign (Empregos pelo clima) and its position as an eco-socialist organization.[46] Nevertheless, neither trade unions nor workers participated in their actions. Equally, Climáximo activists were not present at workers' and trade unions' actions against the unexpected dismissals of workers and the lack of alternatives provided to them. This is despite Climáximo being invited by the Petrogal Central Workers Commission—the main union for workers at the Sines oil refinery—to elaborate collectively a JT plan for the closure of this oil refinery. The organization described this invitation as a sign of the success of their protest against the closure of the refinery, but no collective action developed further between them.

Renewable Energies: Tackling the Climate Emergency Despite Socio-ecological Costs?

The case of activism around renewable energies is partly similar to that of coal-fired power plant and oil refinery closures. Several climate and environmental NGOs have been ambivalently critical regarding the social and labor effects of wind and solar farms. Unlike in the case of closures, some labor and environmental organizations have converged in condemning renewable energy policy. Moreover, labor-environmental alliances have arisen, even though most environmental organizations have partnered with neighbor associations, rather than trade unions.

Since 2020, actions connected to the development of the renewable energy sector have multiplied in Portugal and Spain. Between 2021 and 2023, supreme courts cancelled a large number of renewable energy projects in Andalucía, and especially in Galicia. Behind these decisions were the actions of environmental groups and rural dwellers' associations, often acting through alliances incorporating labor, environmental, and climate dimensions. These struggles coexisted with climate actions that tended to be ambivalently vocal about the negative effects of renewables on humans and the environment.

Galicia has seen the greatest movement against renewable energy projects in Spain. In 2021, it became the Spanish region with the highest number of wind turbines per person.[47] While this meant a strong capacity to mitigate climate change, scientists, environmental organizations, and rural dwellers questioned such growth due to the potential and actual negative effects of wind energy projects on humans and non-humans.

Wind farms have been criticized not only in relation to noise and the destruction of rural dwellers' surrounding environments, but also for their effects on rural labor. In June 2021, Casa Grande de Xanceda, the largest organic producer in Galicia, managed to stop the Gassalla wind farm project from installing twenty-one 200-meter-high turbines at an 800-meter distance away from its dairy farm. Although in February 2023, the project was finally cancelled due to concerns regarding environmental impacts, its implementation would have been disastrous for this company employing 50 people and for its 300 free-range cows, whose well-being was central to the company's identity as an organic producer.[48] In June 2023, in Northern Galicia, four cows were killed by lightning while grazing on land that was part of a wind farm. Following an earlier study confirming that rotating wind turbines can initiate lightning more easily than static objects,[49] the Sindicato Labrego Galego (the main farmers and cattle ranchers' union in Galicia) asked the Galician government to investigate the causes of these deaths and to ensure the preventative suspension of all wind farms occupying lands where cows graze.[50] That same month, the local association "Alarma na Terra de Montes" denounced the rapid expansion of wind farms in Tabeirós-Terra de Montes—an illustrative case of how, without appropriate planning, renewables can have severe impacts on rural labor conditions. This 528.5-km^2 Galician locality is currently home to 13 wind farms with 77 turbines between 180 and 200 meters high.[51] These landscape transformations have multiple effects on cows, among which are significant reductions in pastureland and shade, freedom of movement, access to water, increasing human interference due to the creation of trails for the monitoring of wind turbines, noise, and electromagnetic interference. Local cattle ranchers are already noticing the negative impacts of this decrease in cows' well-being.[52]

Users of commonly owned forests (in Galician *Montes Vecinais*) in Tabeirós-Terra de Montes and other areas of Galicia have mobilized against wind farms due to the effects on their labor conditions, their livelihoods, and the environment. Regarding the latter, they have denounced effects such as the reduced mobility of wild horses, the destruction of groundwaters and natural springs, and the close proximity of wind turbines to areas that are part of Natura 2000 protected area networks.[53]

In February 2023, the Spanish government published its "Plans for the Maritime Space" (POEMs), projecting the installation of approximately 285 wind turbines on the coasts of Galicia, Asturias, and the Basque country. Since then, fishermen and environmental NGOs have questioned the plans. These turbines are more than 230 meters high, and many of them will be installed only 12 meters from the coastline, which in turn will have significant impacts on marine life's well-being and biodiversity.[54]

In this context of rapidly multiplying wind energy projects with negative consequences for humans and non-humans, alliances between rural dwellers' associations, environmental organizations, and associations representing fishermen, cattle ranchers, and farmers have emerged. In November 2020, Adega, Ecologistas en Acción and another 80 neighbor (mostly rural) associations wrote the "Aldeas con Horizonte" (meaning "small villages with horizon") manifesto. This document called for a moratorium of new wind farms in Galicia and the revision of those projects where social conflict around them was taking place. In 2021, the alliance grew to more than 170 collectives from Galicia, forming the "Eólica así non" (meaning "wind energy not this way") platform. This alliance organized demonstrations against the expansion of large-scale renewable energy projects in Galicia after June 5 (World Environment Day) 2021.[55] Today, central to its manifesto, and actively supported by Adega and Ecologistas en Acción,[56] is the defense of rural dwellers' well-being, rural labor, and living conditions, as well as Galician people's ability to decide "on the use of their resources."[57] Galicia's main union, the CIG, has supported actions by this platform, but it has also rejected the closure of wind turbine factories in the region. Moreover, it has represented workers' mobilizations for better employment conditions in wind turbine factories and stood against the closure of these factories in various parts of Galicia.[58]

In Andalusia, the "Stop Parque Eólico Mar de Ágata" manifesto was signed by 217 organizations, including fishermen associations, neighbor associations, environmental organizations, four trade unions, and green parties.[59] The document demanded the cancellation of a 70-km^2 wind farm proposed for installation next to Carboneras and less than 5 km away from the Cabo de Gata-Níjar Natural Park, which contains critical habitats for cetaceans. This labor-environmental alliance was effective, and the project was finally withdrawn.[60] Ecologistas en Acción was part of this coalition. Moreover, one of its local branches in Cádiz province denounced the creation of a solar energy project next to Los Barrios, not only for its closeness to Los Alcornocales Natural park, but also for the "destruction of direct jobs in the primary [farming and cattle ranching] sector, 200 temporary seasonal jobs and many indirect jobs."[61] In contrast, Greenpeace was ambivalently critical regarding the negative effects of renewables on humans and non-human nature, going so far as to defend the renewable energy sector as a source of economic recovery after coal-fired power plant closures:

> Endesa [Carboneras power plant owner] must face its environmental responsibility… giving way to sustainable wind and photovoltaic projects that generate quality jobs and that meet the proclaimed objectives of recovery, transformation and resilience.[62]

However, the implications of these projects for rural dwellers and labor in the fishing, extensive cattle ranching, and agricultural sectors were not clear. Besides, the organization did not sign any of the manifestos against the large-scale wind energy projects mentioned earlier. In March 2023, Manoel Santos, the Greenpeace coordinator for Galicia, critiqued the lack of social dialog and consideration of biodiversity and rural development in on-shore wind energy projects in Spain. At the same time, he suggested that off-shore wind farms could actually benefit ecosystems by excluding activities such as bottom-trawling, cautioning against the negative effects of the fishing sector:

> It is unacceptable to affirm that renewable energies will generate more environmental impacts than a destructive fishery that also emits up to 1,000 million tons per year of carbon dioxide internationally.[63]

These arguments are connected to recent mobilizations by the National Federation of Fishing Associations in Northern Spain against the POEMs, due to their environmental and labor implications. The federation has defended the role of this sector in climate change mitigation as being an animal protein source with fewer CO_2 emissions, but it has failed to acknowledge the need for a JT in this sector to reduce its environmental and climate impacts.[64]

In Portugal, the most visible climate organizations in the context of renewable energy projects have been Zero and Climáximo. In line with Greenpeace, both organizations have been ambivalently critical regarding the negative effects of renewables on labor and ecosystems. In October 2021, Zero called for "a rapid, just and sustainable energy transition [that addresses]… environmental and social objectives at the same time."[65] Similarly, in June 2023, Climáximo highlighted the need for renewable energy development to follow "the deadlines dictated by the climate science." However, like Greenpeace, none of these organizations was clear about how to achieve the expansion of renewables without negatively affecting labor conditions in agriculture, fishing, cattle ranching, rural living environments, and ecosystems.

Sometime after taking these positions, these organizations adopted a more critical perspective. In March 2023, Zero demanded the cancellation of three of the nine proposed off-shore wind projects projected within or next to ecosystems protected by the Natura 2000 network. In October 2023, the organization reiterated that "renewable energy development should not come at a cost for ecosystems and social rights."[66] Notwithstanding this point, this has rarely translated into a critique of specific renewable energy projects due to their consequences on labor and rural living environments. In September 2023, Climáximo denounced the creation of a wind farm in Morgavel that ultimately led to the deforestation of 1,812 cork oak trees. Earlier on, in May 2023, it condemned the Environmental Portuguese Agency's decision to approve a lithium mine in Barroso to supply the renewable energy sector. In contrast, Zero criticized the project but did not make the link between this mine and the development of the renewable energy sector.[67]

Conclusion

This chapter has explored ambivalences around the integration of labor, environmental, and climate goals in mobilizations around energy decarbonization processes in Spain and Portugal. It has illustrated the repeated presence of ambivalences in the discourses and practices of the mobilizing actors studied, as well as its connection with two main features: first, time pressure and limited agency of these actors, due to the climate emergency and the concentration of decision-making power in the hands of energy companies carrying out rapid decarbonization; second, the growing realization among these actors about the labor, social, environmental, and climate implications of the energy sector. The first feature makes such goals work as a zero-sum game, fostering tensions between the defense of labor, the environment, and the climate. In contrast, the second one pushes mobilizing groups toward the integration of environmental, labor, social, and climate issues. The result of combining these two features is constant ambivalence around such integration. Unions and workers affected by coal-fired power plant and oil refinery closures have recognized the need for decarbonization. Nonetheless, the lack of employment alternatives has led them to prioritize job protection, even at the expense of climate change mitigation. This is also the case for rural dwellers and fishermen affected by renewable energy development. Environmental and climate NGOs have recognized the need for a JT in energy decarbonization, but they have prioritized the climate over the labor emergency. Thus, even though some of the studied organizations expressed their interest in cooperating with workers, in practice this did not happen. In the case of large-scale renewable energy projects, some of these organizations have been ambivalently vocal regarding the negative effects of such projects on labor, living conditions in rural areas, and the environment. However, others have focused their actions on the cancellation of large-scale wind farms and solar plants, often in coalition with labor activists and rural dwellers associations. These different responses to energy decarbonization suggest the need to connect ambivalences in socio-ecological transformations to organizations' political trajectories and approaches. At the same time, the evolution of Portuguese climate NGOs in this context suggests, as in the case of unions, that their trajectories are not linear. Further research should shed further light on the processes reshaping activists' integration of labor, social, environmental, and climate goals in ecological and low-carbon transitions. This chapter, however, points to the connection between such integration and organizations' need to reproduce themselves within a context marked by three features: the growing popularization of JTs in climate change policy processes; growing concerns about the climate emergency; and the proliferation of protests by workers, unions, and neighbor associations against energy decarbonization measures.

Finally, the chapter also emphasizes the need to broaden the field of labor environmentalism to include climate struggles, which are not always aligned with environmental ones, as the cases herein have shown. Moreover, such efforts should also involve exploring possibilities around the convergence of workers in sectors affected differently by energy decarbonization. Although experiencing this process during the same period and in close proximity, workers at power plants in Galicia and Andalusia have not collaborated with fishermen and cattle ranchers affected by renewable energy projects. In addition, further research and activism should explore possibilities for dialog between workers at turbine and solar panel factories, i.e., those whose labor and everyday lives are affected by wind energy projects, and climate and environmental activists. While dealing with these contradictions and fragmentations is challenging, it is nevertheless the basis for a broader coalition in the path toward people-centered and environmentally respectful transitions to energy decarbonization.

Notes

1 "2030 Climate Target Plan," *European Commission*, July 14, 2021, accessed May 2, 2023, https://climate.ec.europa.eu/eu-action/european-green-deal/2030-climate-target-plan_en.

2 "Cierre de la centrales térmicas de Carbón en España," *Wikipedia*, last modified March 27, 2024, accessed May 2, 2023, https://es.wikipedia.org/wiki/Cierre_de_las_centrales_térmicas_de_carbón_en_España.

3 "Fecho Central do Pego. Portugal deixa de ter produção de eletricidade a partir do carvão," *RTP*, November 21, 2021, accessed May 20, 2023, www.rtp.pt/noticias/economia/fecho-central-do-pego-portugal-deixa-de-ter-producao-de-eletricidade-a-partir-do-carvao_v1364794; "Central de Sines encerra esta sexta-feira, antes do previsto devido a evolução do Mercado," *Observador*, January 15, 2021, accessed June 1, 2023, https://observador.pt/2021/01/15/central-de-sines-encerra-esta-sexta-feira-antes-do-previsto-devido-a-evolucao-do-mercado/; "Fecho da refinaria de Matosinhos 'continua assombrar' Galp e Governo," *Observador*, July 25, 2022, accessed May 29, 2023, https://observador.pt/2022/07/25/fecho-da-refinaria-de-matosinhos-continua-a-assombrar-galp-e-governo/.

4 "Boletím 2022 Electricidade Renovável," *APREN*, accessed February 15, 2023, www.apren.pt/contents/publicationsreportcarditems/boletim-renovaveis-dezembro-2022.pdf.

5 Ann M. Eisenberg, "Just Transitions," *Southern California Law Review* 92, no. 202 (2008): 273–330.

6 Stefania Barca and Emmanuele Leonardi, "Working-class ecology and union politics: a conceptual topology," *Globalizations* 15, no. 4 (2018): 487–503; Betsy A. Beymer-Farris and Thomas J. Bassett, "The REDD menace: Resurgent protectionism in Tanzania's mangrove forests," *Global Environmental Change* 22 (2012): 332; 517; Amaranta Herrero and Louis Lemkow, "Environmentally Blind Discourses on Coal Extraction and the Idealization of the Miner in Spain," *Capitalism Nature Socialism* 26, no. 4 (2015): 215–235; Tobias Kalt, "Agents of transition or defenders of the status quo? Trade union strategies in green transitions," *Journal of Industrial Relations* 64, no. 4 (2022): 499–521; Adrien Thomas and Nadja Doerflinger, "Trade union strategies on climate change mitigation: Between opposition, hedging and support," *European Journal of Industrial Relations* 26, no. 4 (2020): 383; Rocío Hiraldo and Thomas Tanner, "The Global Political Economy of REDD+: Engaging Social Dimensions in the Emerging Green Economy," *Global Environmental Change* 22 (2012): 217–225.

7 John-Henry Harter, "Environmental Justice for Whom? Class, New Social Movements, and the Environment: A Case Study of Greenpeace Canada, 1971–2000," *Labour/Le Travail* 54 (2004): 83–120. Dimitris Stevis, David Uzzell, and Nora Räthzel, "The labor–nature relationship: varieties of labor environmentalism," *Globalizations* 15, no. 4 (2018): 439.

8 Kenneth Gould, Tammy L. Lewis, and Timmons J. Roberts, "Blue-green coalitions: Constraints and possibilities in the post 9–11 political environment," *Journal of World-Systems Research* X, no. 1 (2004): 91–116; Brian K. Obach, "Labor-environmental relations: An analysis of the relationship between labor unions and environmentalists," *Social Science Quarterly* 83, no. 1 (2002): 82–100.

9 Chrislain Eric Kenfack, "Just Transition at the Intersection of Labour and Climate Justice Movements: Lessons from the Portuguese Climate Jobs Campaign," *Global Labour Journal* 10, no. 3 (2019): 224–239; Sean Sweeney and John Treat, "Trade unions and just transition: The search for a transformative politics," *Trade Unions for Energy Democracy* 11 (April 2018): 1.

10 Gould, Lewis, and Roberts, "Blue-green coalitions," 91; Obach, "Labor-environmental relations," 82.

11 Nora Räthzel, Dimitris Stevis, and David Uzzell, "Introduction: Expanding the Boundaries of Environmental Labor Studies," in *The Palgrave Handbook of Environmental Labor Studies*, eds. Nora Räthzel, Dimitris Stevis and David Uzzell

(Cham, SUI: Palgrave Macmillan, 2021), 3. Joan Martinez Alier, *The Environmentalism of the poor. A study of ecological conflicts and evaluation* (Northampton, MA: Edward Elgar, 2002), 1–10; Nixon Rob, *Slow Violence and the Environmentalism of the Poor* (Cambridge, MA: Harvard UP, 2011), 353. Stefania Barca, "On working-class environmentalism: A historical and transnational over-view," *Interface* 4, no. 2 (2012): 61; Karen Bell, *Working-Class Environmentalism: An Agenda for a Just and Fair Transition to Sustainability* (Cham, SUI: Palgrave Macmillan, 2020), 139–141; Robert D. Bullard, *Dumping in Dixie: Race, Class, and Environmental Quality* (Boulder: Westview Press, 2000), 137–140; Lara Domínguez and Colin Luoma, "Decolonising Conservation Policy: How Colonial Land and Conservation Ideologies Persist and Perpetuate Indigenous Injustices at the Expense of the Environment," *Land* 9, no. 65 (January 2020): 1–22. Steve Lerner, *Sacrifice Zones: The Frontlines of Toxic Chemical Exposure in the United States* (Cambridge: MIT Press, 2010).

12 Barca and Leonardi, "Working-class ecology," 487; Todd Vachon, "The Green New Deal and Just Transition Frames within the American Labor Movement," in *The Palgrave Handbook of Environmental Labor Studies*, eds. Nora Räthzel, Dimitris Stevis and David Uzzell (Cham, SUI: Palgrave Macmillan, 2021), 105; Herrero and Lemkow, "Environmentally Blind Discourses on Coal Extraction," 215.

13 Kalt, "Agents of transition or defenders of the status quo?" 499; Obach, "Labor-environmental relations," 86. Dimitris Stevis, "Embedding Just Transition in the USA: The Long Ambivalence," in *The Palgrave Handbook of Environmental Labor Studies*, eds. Nora Räthzel, Dimitris Stevis and David Uzzell (Cham, SUI: Palgrave Macmillan, 2021): 591–619.

14 Nora Räthzel, "Individuals Transforming Organisations: Spanish Environmental Policies in Comisiones Obreras," in *The Palgrave Handbook of Environmental Labor Studies*, eds. Nora Räthzel, Dimitris Stevis and David Uzzell (Cham, SUI: Palgrave Macmillan, 2021): 647–674.

15 Dimitris Stevis and Romain Felli, "Global labor unions and Just Transition to a green economy," *International Environmental Agreements* 15 (2015): 29–43; Sweeney and Treat, "Trade unions and just transition."

16 Hadrien Malier, "No sociological excuses for not going Green: how do environmental activists: how do environmental activists make sense of social inequalities and relate to the working class," *European Journal of Social Theory* 24, no. 3 (2021): 411–430.

17 Beymer-Farris and Bassett, "The REDD menace"; Harter, "Environmental Justice for Whom?" 83–120; Hiraldo and Tanner, "The Global Political Economy of REDD+," 7; Becca Wilgosh, Alevgul H. Sorman, and Iñaki Barcena, "When two movements collide: Learning from labor and environmental struggles for future Just Transitions," *Futures* 137 (2022): 102903.

18 Gould, Lewis, and Roberts, "Blue-green coalitions."

19 Erik Loomis, "Working-classs environmentalism: the case of Northwest Timber Workers," in *The Palgrave Handbook of Environmental Labor Studies*, eds. Nora Räthzel, Dimitris Stevis and David Uzzell (Cham, SUI: Palgrave Macmillan, 2021), 127.

20 Halliki Kreinin, "Typologies of 'Just Transitions': Towards Social-Ecological Transformation," *Ecological Economic Papers* 35 (2020): 1; Stevis and Felli, "Global labor unions," 29; Wilgosh, Sorman, and Barcena, "When two movements collide," 102903.

21 "Just Transition Fund," *European Commission*, accessed May 7, 2022, https://ec.europa.eu/regional_policy/funding/just-transition-fund_en.

22 "Central de Sines encerra esta sexta-feira," *Observador*; "Câmara de Abrantes alerta para questões sociais e económicas com fecho da central do Pego," *Observador*, November 22, 2021, accessed May 24, 2023, https://observador.pt/2021/11/22/camara-de-abrantes-alerta-para-questoes-sociais-e-economicas-com-fecho-da-central-do-pego/; "Galp avança com despedimento coletivo de 150 trabalhadores de Matosinhos," *Publico*, May 6, 2021, accessed May 18, 2023, www.publico.pt/2021/05/06/economia/noticia/galp-avanca-despedimento-coletivo-150-trabalhadores-matosinhos-1961540.

23 The Just Transition Strategy drafted by the government was ambiguous with how auxiliary workers should be supported, hence leaving the opportunity for energy companies to neglect them in the provision of alternatives after the closure of the plant.

24 "Los trabajadores auxiliares de Endesa se movilizan con un corte de tráfico," *La Voz de Galicia*, January 14, 2021, accessed May 3, 2023, www.lavozdegalicia.es/noticia/ferrol/as-pontes-de-garcia-rodriguez/2021/01/14/trabajadores-auxiliares-endesa-movilizan-corte-trafico/00031610633581570348939.htm; "Camioneros y sindicatos bloquean la central térmica de As Pontes," *La Voz de Galicia*, June 25, 2020, accessed May 18, 2023, www.lavozdegalicia.es/noticia/ferrol/as-pontes-de-garcia-rodriguez/2020/06/25/camioneros-sindicatos-bloquean-central-termica-as-pontes/00031593086438889145739.htm; "Entrevista con trabajador de Carboneras," *emartv*, June 11, 2021, accessed May 5, 2023, www.emartv.com/ftpentrevistas/central-termica-carboneras-cierre-210611.mp3; "Despedidos de DITECSA piden recuperar sus empleos con una marcha en Los Barrios," *Andalucía información*, September 11, 2020, accessed May 28, 2023, https://andaluciainformacion.es/campo-de-gibraltar/919079/despedidos-de-ditecsa-piden-recuperar-sus-empleos-con-una-marcha-en-los-barrios/; "Trabalhadores da Petrogal em manifestação contra o fecho da refinaria de Matosinhos," *Observador*, February 2, 2021, accessed May 15, 2023, https://observador.pt/2021/02/02/trabalhadores-da-petrogal-em-manifestacao-contra-o-fecho-da-refinaria-de-matosinhos/.

25 "La central Térmica de Puente Nuevo teme por el futuro de sus trabajadores," *Córdoba buenas noticias*, December 5, 2019, accessed February 20, 2023, www.cordobabn.com/articulo/cordoba/la-central-termica-de-puente-nuevo-teme-por-el-futuro-de-sus-trabajadores/20191205135354025285.html; "Empleados de Meirama denuncian la velocidad de los despidos sin alternativa," *El Correo Gallego*, June 18, 2020, accessed April 20, 2023, www.elcorreogallego.es/galicia/empleados-de-la-central-de-meirama-denuncian-la-velocidad-de-los-despidos-sin-alternativas-AE3547662.

26 "Transición enerxética galega e xusta," *Confederación Intersindical Galega*, accessed May 20, 2023, www.cig.gal/propostas-saida-crise-transicion-enerxetica.html; "Transição energética justa em debate internacional amanhã," *CGTP*, February 24, 2022, accessed May 20, 2023, www.cgtp.pt/accao-e-luta/17213-transicao-energetica-justa-em-debate-internacional-amanha.

27 "Sindicato: fecho da central de Sines deve ser repensado," *Diário do Alentejo*, accessed February 21, 2023, https://diariodoalentejo.pt/pt/noticias/8729/sindicato-fecho-da-central-de-sines-deve-ser-"repensado.aspx; "Encerramentos desastrosos não são transição energética justa," *CGTP*, November 18, 2021, accessed May 28, 2022, www.cgtp.pt/accao-e-luta-geral/16872-encerramentos-desastrosos-nao-sao-transicao-energetica.

28 "La CIG solicita la reactivación de dos de los cuatro grupos de la central térmica de As Pontes," *Diario de Ferrol*, July 15, 2022, accessed February 27, 2023, www.diariodeferrol.com/articulo/comarcas/cig-solicita-reactivacion-completa-cuatro-grupos-central-termica-as-pontes-3827050.

29 "Despedidos," *Andalucía información.*

30 Quote translated by the author, "UGT recomienda a los exempleados de la térmica que no se reincorporen a sus puestos de trabajo," *UGT Viesgo Fica*, March 25, 2022,

accessed May 18, 2023, https://viesgo.ugt-fica.org/component/content/article/36-inicio/1197-ugt-recomienda-a-los-exempleados-de-la-termica-que-no-se-reincorporen-a-sus-puestos-de-trabajo.

31 "El cierre de la central de Carboneras supondrá un gran paso en la lucha contra el cambio climático," *Greenpeace*, October 4, 2019, accessed May 21, 2023, https://es.greenpeace.org/es/sala-de-prensa/comunicados/el-cierre-de-la-central-de-carboneras-supondra-un-gran-paso-en-la-lucha-contra-el-cambio-climatico/; "Cierre de la central térmica de carbón litoral de Almería," *Ecologistas en acción*, October 2, 2019, accessed May 22, 2023, www.ecologistasenaccion.org/128041/cierre-de-la-central-termica-de-carbon-litoral-de-almeria/;"El hidrógeno verde como salvavidas de la industria del gas," *Ecologistas en acción*, October 21, 2022, accessed May 9, 2023, www.ecologistasenaccion.org/211105/el-hidrogeno-verde-como-salvavidas-de-la-industria-del-gas/.

32 "La central Térmica de Puente Nuevo," *Córdoba buenas noticias*.; "Los trabajadores de la central de Meirama harán huelga indefinida a partir de este martes," *Europa press*, March 9, 2020, accessed May 13, 2022, www.europapress.es/galicia/noticia-trabajadores-central-meirama-haran-huelga-indefinida-partir-martes-20200309180537.html; "A CIG denuncia a falta de compromisos firmes por parte de Endesa e do Goberno no mantemento do emprego tras o peche da térmica das Pontes," *Confederación Intersindical Galega*, September 7, 2023, accessed October 3, 2023, www.cig.gal/nova/a-cig-denuncia-a-falta-de-compromisos-firmes-por-parte-de-endesa-e-do-goberno-no-mantemento-do-emprego-tras-o-peche-da-termica-d.html; "Fundo de Transição vai ter 40 a 45 milhões para central do Pego," *Observador*, November 13, 2021, accessed May 21, 2022, https://observador.pt/2021/11/13/fundo-de-transicao-vai-ter-40-a-45-milhoes-para-central-do-pego/.

33 Ana Delgado and Kjetil Rommetveit, "Our strength is diversity: imaginaries of nature and community in a Brazilian social movement," *International Journal of Sustainable Development* 15, no. 4 (2012): 353–373.

34 "El lado oscuro del Carbón," *Greenpeace*, September 27, 2018, accessed May 20, 2022, www.youtube.com/watch?v=OyCySQ7SnPw.

35 "Apaga el carbón," *Ecologistas en acción*, November 27, 2018, accessed May 10, 2022, www.ecologistasenaccion.org/110358/apaga-el-carbon/.

36 "Único Sentido. 2025 Sin Carbón ni Energía Nuclear," *Greenpeace*, March 2018, https://es.greenpeace.org/es/wp-content/uploads/sites/3/2018/03/RESUMEN-EJECUTIVO-LR.pdf.

37 "Carbón," *Greenpeace*, accessed May 30, 2023, https://es.greenpeace.org/es/trabajamos-en/cambio-climatico/carbon/.

38 "Galiza. Un futuro sen carbon," *Un futuro sen carbón*, November 2018, accessed May 18, 2023, https://unfuturosencarbon.org/manifesto/.

39 "Lanzamiento de la campaña Europa un futuro sin carbón," *Ecologistas en acción*, November 30, 2018, accessed May 12, 2023, www.ecologistasenaccion.org/110789/lanzamiento-de-la-campana-europa-un-futuro-sin-carbon/.

40 "Quen somos," *Adega*, accessed March 31, 2023, https://adega.gal/info.php?sec=3&idioma=gl; "¿Quiénes somos?" *Ecologistas en acción*, accessed March 31, 2023, www.ecologistasenaccion.org/quienes-somos/.

41 "Desactivada definitivamente una de las centrales térmicas más contaminantes del país," *Ecologistas en acción*, December 20, 2021, accessed April 3, 2023, www.ecologistasenaccion.org/186503/desactivada-definitivamente-una-de-las-centrales-termicas-mas-contaminantes-del-pais/; "Greenpeace y Ecologistas en Acción rechazan cualquier pretensión de retrasar el cierre de la térmica de Carboneras (Almería)," *Greenpeace*, May

14, 2021, accessed May 18, 2022, https://es.greenpeace.org/es/sala-de-prensa/comunicados/greenpeace-y-ecologistas-en-accion-rechazan-cualquier-pretension-de-retrasar-el-cierre-de-la-termica-de-carboneras-almeria/; "Hoxe pecha a central de Meirama, tras 40 anos contaminando nostra terra," *Adega*, accessed June 1, 2023, https://adega.gal/web/novas.php?sec=209&id=998&idioma=g.

42 Quote translated by the author. "Verdemar denuncia los primeros vertidos de la térmica de los Barrios tras su reapertura," *Initiativa Comarcal*, January 24, 2022, accessed May 20, 2023, https://iniciativacomarcal.com/verdemar-ecologistas-denuncian-los-primeros-vertidos-de-carbon-de-la-termica-de-los-barrios-tras-su-reapertura/.

43 "Lanzamiento," *Ecologistas en acción.*

44 "Fecho da central de carvão de Sines-Zero congratula-se mas quer melhor planeamento na transição energética," *Zero*, January 14, 2021, accessed May 1, 2023, https://zero.ong/noticias/fecho-da-central-de-carvao-de-sines-zero-congratula-se-mas-quer-melhor-planeamento-na-transicao-energetica/; "Portugal deixou definitivamente de usar carvão na produção de eletricidade," *Zero*, November 21, 2021, accessed May 2, 2023, https://zero.ong/noticias/portugal-deixou-definitivamente-de-usar-carvao-na-producao-de-eletricidade-pego-era-responsavel-por-4-das-emissoes-do-pais/.

45 Quote translated by the author. "Qual foi a liçao que Costa deu à Galp?" *Climaximo*, April 19, 2023, accessed May 2, 2023, www.climaximo.pt/2023/04/19/qual-foi-a-licao-que-costa-deu-a-galp-leonor-canadas/.

46 "Etiqueta: ecossocialismo," *Climaximo*, accessed February 20, 2023, www.climaximo.pt/tag/ecossocialismo/.

47 "Manifestación 5 Xuño; Eólica, Así Non!," *Verdegaia*, May 29, 2021, accessed May 20, 2023, www.verdegaia.org/wp/manifestacion-5-de-xuno-eolica-asi-non/.

48 "Un macroproyecto eólico amenaza al mayor productor ecológico de Galicia," *Economía Digital Galicia*, August 30, 2021, accessed May 20, 2023, www.economiadigital.es/galicia/empresas/un-macroproyecto-eolico-amenaza-al-mayor-productor-ecologico-de-galicia.html; "La amenaza para Xanceda del macroparque Ordes-Curtis," *Casa grande de Xanceda Ecológico*, May 21, 2021, accessed May 20, 2023, www.casagrandexanceda.com/ampliar/707/La-amenaza-para-Xanceda-del-macroparque-Ordes-Curtis.html.

49 Joan Montanyà, Oscar van der Velde, and Earle R. Williams, "Lightning discharges produced by wind turbines," *Journal of Geophysical Research: Atmospheres* 119 (2014): 1455–1462.

50 "Inquietud por la muerte de cuatro vacas por un rayo al pie de un parque eólico," *Campo Galego*, June 15, 2023, accessed July 1, 2023, www.campogalego.es/inquietud-por-la-muerte-de-4-vacas-por-un-rayo-al-pie-de-un-parque-eolico/.

51 "Trece parques con 77 aerogeneradores de casi 200 metros afectan a la zona," *La Voz de Galicia*, June 14, 2023, accessed July 2, 2023, www.lavozdegalicia.es/noticia/deza/2023/06/14/trece-parques-77-aerogeneradores-200-metros-afectan-zona/0003_202306D14C5992.htm

52 Interview with representative of Alarma na Terra de Montes, March 20, 2023.

53 "La Comunidad de Montes de Viladesuso rechaza la construcción de Parques eólicos en su monte," *Telemariñas*, February 19, 2022, accessed July 3, 2023, https://telemarinas.com/la-comunidad-de-montes-de-viladesuso-rechaza-la-construccion-de-parques-eolicos-en-su-monte/.

54 Josep Lloret et al., "Unravelling the ecological impacts of large-scale offshore wind farms in the Mediterranean Sea," *Science of The Total Environment* 824 (2022): 153803.

55 "Más de 200 colectivos se manifiestan en Santiago contra los nuevos parques eólicos en proyecto," *La Voz de Galicia*, June 5, 2022, accessed May 20, 2023, www.lavozdegalicia.es/noticia/santiago/santiago/2022/06/05/200-colectivos-manifiestan-santiago-contra-nuevos-parques-eolicos-proyecto/00031654430052449768585.htm.
56 "A Coordinadora 'Eólica Así Non' chama o pobo galego a rodear a Xunta o domingo 23 de abril," *Adega*, March 22, 2023, accessed May 24, 2023, https://adega.gal/novas.php?id=1352&idioma=gl&sec=7; "Gran Manifestación en Compostela 'Eólica Así Non'," *Ecologistas en acción*, June 5, 2022, accessed September 2, 2022, www.ecologistasenaccion.org/201207/gran-manifestacion-en-compostela-eolica-asi-non/.
57 "Gran Marcha no Día de Medio Ambiente," *Verdegaia*, June 2021, accessed May 23, 2023, www.verdegaia.org/wp/wp-content/uploads/2021/05/Manifesto-EOLICA-ASI-NON.pdf.
58 "A CIG rexeita o peche da fábrica de Vestas e insta a Xunta a adoptar as medidas necesarias para evitar os despedimentos," *Confederación Intersindical Galega*, September 22, 2021, accessed October 2, 2023, www.cig.gal/nova/a-cig-rexeita-o-peche-da-fabrica-de-vestas-e-insta-a-xunta-a-adoptar-as-medidas-necesarias-para-evitar-os-despedimentos.html; interview with employee at wind turbine factory in Ferrol, March 30, 2023; interview with CIG representative, March 20, 2023.
59 "Manifiesto 'No al Parque Eólico Mar de Ágata'," accessed March 30, 2023, https://stopparqueeolicomardeagata.org/manifiesto-contra-parque-eolico-mar-de-agata-cabo-de-gata/
60 "Los promotores del parque eólico marino 'Mar de Ágata' frente a Cabo de Gata (Almería) abandonan el proyecto," *Europa Press*, March 1, 2023, accessed May 5, 2023, www.europapress.es/andalucia/almeria-00350/noticia-promotores-parque-eolico-marino-mar-agata-frente-cabo-gata-almeria-abandonan-proyecto-20230301144849.html.
61 "Oposición a la gran fotovoltaica de Castellar y San Roque," *Ecologistas en acción*, April 17, 2021, accessed February 4, 2023, www.ecologistasenaccion.org/168240/oposicion-a-la-gran-fotovoltaica-de-castellas-y-san-roque/.
62 "Greenpeace y Ecologistas," *Greenpeace*.
63 "Eólica Marina. Prioricemos los usos sostenibles del mar." El Diario.es, accessed May 2, 2023, www.eldiario.es/galicia/blog/opinion/eolica-marina-prioricemos-usos-sostenibles-mar_132_10003112.html
64 "Manifesto de Burela en Defensa del Sector Pesquero del Cantábrico – Noroeste," *FNCP*, November 4, 2021, accessed March 13, 2023, www.fncp.eu/wp-content/uploads/2021/11/MANIFESTO-DE-BURELA-EN-DEFENSA-DEL-SECTOR-PESQUERO-DEL-CANTABRICO-—-NOROESTE.pdf.
65 "Juntos por uma Europa 100% Renovável," *Zero*, accessed May 12, 2023, https://zero.ong/noticias/juntos-por-uma-europa-100-renovavel/ "Manifesto por um serviço público de energias renováveis," *Climaximo*, May 24, 2023, accessed May 25, 2023, www.climaximo.pt/2023/05/24/manifesto-por-um-servico-publico-de-energias-renovaveis/.
66 "Proposta energia eólica offshore: ONGA propoem criterios ambientais e socioeconómicos mais rigorosos," *Zero*, March 9, 2023, accessed May 2, 2023, https://zero.ong/noticias/proposta-energia-eolica-offshore-onga-propoem-criterios-ambientais-e-socioeconomicos-mais-rigorosos/; "ZERO entende que uma aplicação inteligente e ponderada da Diretiva Energias Renováveis pode acelerar o desenvolvimento das energias renováveis em harmonia com o ambiente e o tecido social," *Zero*, October 9, 2023, accessed November 1, 2023, https://zero.ong/noticias/50-organizacoes-da-sociedade-civil-apelam-a-compatibilidade-da-aceleracao-das-energias-renovaveis-com-a-natureza-e-as-populacoes/.

67 "O movimento pela justiça climática diz não aos projetos de mineração de lítio em Portugal: porquê?" *Climaximo*, May 31, 2023, accessed October 2, 2023,https://arquivo.climaximo.pt/2023/05/31/o-movimento-pela-justica-climatica-diz-nao-aos-projetos-de-mineracao-de-litio-em-portugal-porque/; "Cortar sobreiros corta emissões?-Leonor Canadas," *Climaximo*, September 1, 2023, accessed October 2, 2023, https://arquivo.climaximo.pt/2023/09/01/cortar-sobreiros-corta-emissoes-leonor-canadas/; "ZERO alerta para 6 pontos negativos na exploração de lítio no Barroso," *Zero*, April 19, 2023, accessed October 2, 2023, https://zero.ong/noticias/zero-alerta-para-6-pontos-negativos-na-exploracao-de-litio-no-barroso/.

12 Bioplastics versus Conventional Plastics

An Analysis from a Sociological, Ethical, and Educational Perspective

Ulrike Ohl, Maria Backhouse, Kerstin Schlögl-Flierl, Anja Kalch, and Helena Bilandzic

Introduction

Humanity—especially in the Global North—produces enormous amounts of plastic waste. Between 1950 and 2019, plastic waste globally has grown from 2 million to 6.3 million tons annually.[1] Plastics are a socio-ecological problem because they take several decades, or even centuries, to decompose, depending on their chemical composition. In addition, they are produced using crude oil and therefore contribute not only to the overuse of raw materials and the pollution of the earth, but also to climate change.[2] Images of plastic islands in the sea and dead birds with stomachs full of plastics are omnipresent in the media, and researchers are now detecting microplastics not only in water, soils, and fish, but also in human bodies.[3]

One solution is to reduce the individual consumption of products, but there are limits to this strategy because consumers, through their purchasing behavior, can only influence certain parts of the plastics economy, i.e., in many sectors, from car manufacturing to agriculture, plastics cannot be easily eliminated. From food to textiles and the medical sector, plastic materials have become entrenched,[4] so it is not surprising that global plastics consumption continues to increase despite knowledge of their harmful impacts.

An important policy in this regard has been the development and expansion, since the 1990s, of bioplastics.[5] Unlike conventional plastics, bioplastics are not based on fossil resources but are entirely or partly made from plant biomass (e.g., wood, corn, sugarcane residues or animal products such as chitin). The term "bioplastics" is not protected and encompasses both materials that decompose faster than fossil plastics and biomass-based materials that do not differ chemically from conventional plastics and are therefore not necessarily biodegradable. The argument of lobbyists such as *European Bioplastics* is that they not only offer a way out of dependence on raw fossil materials, but also contribute to climate protection, because they are based on biomass instead of crude oil, and their carbon footprint would be better than that of conventional plastics.[6]

Currently, various EU and German programs fund research on bioplastics, mainly within the frame of transforming to a bioeconomy or a circular economy. The aim of the former is to move society and the economy away from non-renewable resources to renewable biological resources,[7] thus realizing a circular

DOI: 10.4324/9781032627984-15

economy: "In a circular economy, the use of non-renewable resources and waste production is minimized, while reuse and recycling dominate the life cycles of materials."[8]

Despite impressive growth in usage, bioplastics currently do not play an important role in the plastics sector, and according to *European Bioplastics*, "[they] still represent less than one percent of the more than 390 million tons of plastic produced annually" on a global scale.[9] The association expects that global bioplastics production capacities will increase about threefold by 2027 (from 2.23 million tons in 2022 to approximately 6.3 million tons in 2027).[10]

However, there are also conflicting goals regarding bioplastics, in that associated ambivalences make it difficult for consumers and decision-makers to recognize whether their production and consumption represent an adequate approach to socio-ecological transformation. In this chapter, we therefore shed light on the ambivalences connected with bioplastics from sociological, ethical, and educational perspectives, because we assume that transparent information is crucial in the context of a socio-ecological transformation policy that will allow consumers and decision-makers to critically weigh up the advantages and disadvantages of bioplastics for themselves.

First, in the section "Ambivalences of Bioeconomic Policies from a Sociological Perspective", we introduce bioplastics as part of broader socio-ecological transformation strategies, such as the bioeconomy and the circular economy, and outline their ambivalences from a sociological perspective. Then, in the section "Ambivalences from an Ethical Perspective", we discuss these ambivalences from an ethical perspective and debate strategies to deal with the challenge that there are no easy answers to the question relating to the benefits of and harm caused by bioplastics. Since education is an important factor in science communication, i.e., knowledge and competencies can be conveyed that are important in the context of a socio-ecological transformation, in the section "Implications for Educational Approaches to Bioplastics", we derive implications for promising educational approaches on the topic of bioplastics in the sense of an emancipatory education for sustainability. We see educational materials as pivotal in a transparent transformation policy. Therefore, in the section "Explorative Case Study: Analysis of Teaching Materials for Primary Schools on the Topic of Bioplastics", we use the case study of educational material on bioplastics published by the *German Federal Ministry of the Environment, Nature Conservation, Nuclear Safety and Consumer Protection* (BMUV). In this regard, we examine the extent to which the identified ambivalences are addressed therein and the material thus realizes an emancipatory approach to education for sustainability. In the final section, we summarize implications for the political debate on bioplastics.

Ambivalences of Bioeconomic Policies from a Sociological Perspective

The socio-ecological transformation strategy toward a bioeconomy was first introduced by the OECD in 2009 and adopted by numerous countries and the EU in the following years.[11] Germany launched its first strategy in 2010[12] and its second in 2020.[13]

There is no single definition of what constitutes a "bioeconomy." Depending on the context, countries focus more on the promotion of biotechnologies, renewable resources (biomass) or a circular economy. However, all strategies share the technological optimism of ecological modernization, namely that the ecological crisis must be dealt with by means of technological innovations. The central idea behind ecological modernization is that economic growth can be decoupled from the overuse of natural resources by means of green technologies.[14] The drivers of this transformation strategy are supposed to be influential institutions and companies, while the direction of this transformation is to be set by scientific expert fora.[15]

This strategy has been criticized since the very beginning for its technological optimism.[16] Critics problematize that power imbalances and social inequalities which permeate also technological solutions as well as the global supply chains of raw materials are ignored by decision-makers. The bioeconomy thus threatens to exacerbate global socio-ecological problems.[17] Socio-ecological transformation would have to initiate structural change of the "mode of production and living"—a concept which was coined by Ulrich Brand and Markus Wissen and describes the dominant capitalist way of production and consumption.[18]

In the meantime, studies have illustrated in more detail the ambivalences associated with bioeconomy policies. These include the additional demand for biomass (e.g., for bioplastics or bioenergy) and the resulting cultivation and processing of required renewable raw materials that may exacerbate deforestation, lead to overusing fresh water, eutrophication, acidification and the over-fertilization of soils and waters. In the long term, the bioeconomy could therefore cause greater burdens in other environmental areas such as the climate or lead to biodiversity loss.[19]

Linked to this notion, social inequalities such as unequal access to land may increase. Case studies on the soy, sugarcane, and palm oil sectors underline that biomass production has caused social conflicts over land use, expulsions, and precarious working conditions in production regions for decades—and in some cases, centuries.[20] Thus, the bioeconomy currently reproduces existing extractive relations and unequal exchanges between raw material-producing and -processing countries.[21]

Finally, the objectives of the bioeconomy have not yet been the subject of a broad democratic negotiation process (neither in Germany nor in other countries) but are mainly decided in expert fora. In Argentina, for example, the orientation of the bioeconomy is mainly defined by the agricultural and bioenergy sectors,[22] all of which explains why knowledge and innovation in the agricultural sector are so strongly geared to the needs of agribusiness.

Bioplastics reflect these bioeconomic socio-ecological ambivalences, since the production of plant-based raw materials usually takes place in the agro-industrial sector, which contributes substantially to the emission of greenhouse gases due to pesticide use and land conversion.[23] If this is factored in, the carbon footprint of bioplastics can be just as negative as that of fossil plastics.[24]

Furthermore, residual materials from agriculture or forestry are also not infinite, and plant materials from the food production sector are used:

> In particular, land consumption, i.e. the use of agricultural land for production, is one of the biggest criticisms and obstacles for bioplastics. The global area that can be used for agriculture is limited and is used to grow food and feed as well as plants for material use.[25]

Thus, a competitive situation between food and fuel (the "food-versus-fuel debate") will arise due to food or other farming products forming the basis for bioplastic production—corn is an example in this regard. In terms of resource consumption, therefore, only a shift, and no solution, is visible, which is what we discuss ethically in the following.

However, the negative impacts go beyond the agricultural sector.[26] First, like conventional plastics, bioplastics consume a lot of energy during production, regardless of whether they are made solely from plant-based raw materials or are biodegradable. This means that they do not necessarily have a better energy balance than conventional plastics. Second, global inequalities in material flows (e.g., for the production of plastics), as well as the negative impacts of plastic waste, which nowadays is very often exported to marginalized regions and communities in Africa or Asia, are perpetuated. Third, bioplastics can be harmful to health, due to their chemical composition.[27] Finally, the lack of recycling possibilities must also be noted. As already mentioned, some bioplastics can only be disposed of and not composted, which is what the word "organic" suggests. Besides, the chemical composition of compostable plastics can also be harmful, which is why many German cities now ban bags made from bioplastics being thrown into the organic waste garbage bin.[28]

Furthermore, even 20 years after their introduction, bioplastics in Germany are not sufficiently labeled, and consumers are misled by the prefix "bio," which is not protected for bioplastics, unlike for organic food, and thus does not necessarily indicate biologically safe and sustainable materials. Thus, the consumer service in southern Germany recommends generally consuming less plastic and using it for as long as possible.[29]

According to Behnsen und Endres,[30] the public bioplastic discourse is characterized by a very emotionally driven debate with a lack of information flow from science and insufficient information policy. In the same vein, Blesin and Möhring also criticize information deficits:

> Bioplastics are positively evaluated by consumers [...]. However, this positive evaluation was also based on erroneous assumptions and generalization, for example, that a renewable raw material base is per se "better" (more sustainable) than a fossil one [...]. Positively connoted associations with organic food and corresponding attributes such as organic and ecological are assumed on the consumer side.[31]

As these statements remind us, communicative transparency is a crucial prerequisite when it comes to dealing in constructive ways with the ambivalences described earlier. Taking this issue seriously, in the following section we outline an ethical analysis of bioplastics and their ambivalences in the field of socio-ecological transformation and determine why citizens need to be enabled to make their own decisions. Despite all the negative aspects of bioplastics, it should also be emphasized that rethinking how we consume plastics is an important step.

Ambivalences from an Ethical Perspective

As the consideration of sustainability aspects shows, the verdict on bioplastics is rather mixed and not particularly clear-cut: "It should be noted that bioplastics and products made from them cannot per se be described as more sustainable or ecologically more sensible alternatives to conventional plastics."[32] Bioplastics therefore appear to be the lesser of two evils compared to conventional plastics, but on closer inspection they seem to make false promises or only half deliver on promises, so they are not the advertised alternative. As such, there remains a (bio) plastic problem. The ambivalences inherent in bioplastics in particular, as well as in the entire bioeconomy strategy, make this classification clear.

Of course, various adjustments can still be made, but the fundamental question is whether the goal should not be to eliminate plastics as far as possible for reasons of sufficiency, in which case the general reduction of plastics—and, ultimately, avoidance—would be the goal and not an alternative strategy to bioplastics.

One can distinguish between a norm-ethical and a virtue-ethical approach in this field. A norm-ethical approach would choose the path of certain bans, which is, for example, how plastic straw hulls were banned throughout the EU in 2021. In this regard, the norm "avoidance of plastics" was cast into law. At first glance, this seems to have been a sensible and lean measure; however, it was not considered that plastic straws do actually have important functions in various areas, such as the care sector.

With a norm-ethical approach, any exceptions must always be reconsidered. From this perspective, one could also argue that bioplastics should be banned completely, given their various disadvantages and false promises, but this would not serve in terms of innovation or possible improvements to bioplastics. Consequently, the most that can be said is that the use of bioplastics should be handled very restrictively.

A virtue-ethical approach, which focuses on the individual and his or her way of acting, takes a completely different approach. Ever since Aristotle, virtues have been understood as a disposition to be practiced. Whereas in antiquity and the Middle Ages, justice and moderation were seen as virtues in terms of their meaning, in our time, for example, the virtue of temperance, in this special case plastic fasting,[33] could be brought forward. This would then be understood as an everyday practice of thinking about the necessity of plastic every time we come into contact with it, whether we are shopping or working. In this scenario, plastic would not be completely abolished but reduced to a minimum.

Nevertheless, this means that it would be very much up to the individual and that they would have to decide and weigh this up again and again. In contrast, a norm-ethical approach does not usually raise awareness, and critical-emancipative thinking cannot be achieved if regulation only takes place at the level of prohibition and when underlying rethinking is forgotten. In terms of norm ethics, the individual would carry out the thing to be done without necessarily understanding the reason behind it. In the best case, of course, norms are communicated in such a way that everyone can understand them. However, with a topic like plastics, the individual contribution should not be underestimated. Ethically, this would fall within the realm of consumer responsibility. In recent discussions, there are constant warnings of this individualization trap and putting too much of a burden on the individual. The growth paradigm of society is also viewed judiciously in this critical-emancipative view.

The consequence of the growth paradigm is an increase in plastics consumption, even with the help of bioplastics, as shown herein. The substitution of bioplastics, which also should be considered from the point of view of virtue ethics, can certainly be helpful and purposeful in appropriate individual cases. However, it does not aim at a turnaround in the growth paradigm, as the socio-ecological transformation would do; instead, an apparently more ecological variant facilitates the decision to continue using plastics.

From an ethical perspective, norm-ethical and virtue-ethical approaches must be intertwined against the backdrop of the discussion. The individual and the structure must therefore go together in reducing plastics consumption. A purely norm-ethical approach does not bring about a change in thinking, and a purely virtue-ethical approach also needs the appropriate framework conditions so that the individual and the respective action can be effective.

Education plays an important role in making an emancipated ethical judgment and raising awareness so that framework conditions can be adapted and socio-ecological transformation accelerated. Therefore, the following section highlights implications for educational approaches to bioplastics in the context of an emancipatory education for sustainability.

Implications for Educational Approaches to Bioplastics

As the previous reflections reveal, the question of how to deal appropriately with the plastic problem is highly relevant in the context of the socio-ecological transformation of society. The fact that we are constantly confronted with plastics in our everyday life, and from our early childhood, makes the plastics issue at the same time a very instructive topic in schools. The topic is already covered in Germany, for example, in the primary school subject *Sachunterricht* (which addresses topics from the disciplines of geography, biology, physics, and chemistry) and more generally in the secondary school subjects geography, biology, physics, and chemistry.

Since the plastic problem raises questions about sustainable action, and because it requires ecological, social, economic, and political considerations, it is a characteristic topic of the interdisciplinary approach of education for sustainability.[34] Modern emancipatory methods in this regard, which are currently also discussed under the term "transformative learning," explicitly address the ambivalences surrounding socio-ecological transformation in the classroom.[35]

Transformative learning strives for educational processes in which uncertainties regarding the "right paths" to a sustainable future, as well as contradictions relating to sustainable action, are specifically addressed. Existing habits and world views need to be reflected more fundamentally against the backdrop of a complex, uncertain, and insecure but shapeable future.[36] Behind this notion is an understanding of education as a

> process of emancipation, i.e., the promotion of one's thinking and acting based on reflection and the deconstruction of prevailing power relations and narratives, to be able to stand up for one's convictions and realize them through participation in democratic negotiation processes.[37]
>
> (Translated by the authors)

To deal with bioplastics in school in the sense of transformative learning means specifically highlighting their advantages and disadvantages as well as associated subject-specific ambivalences and ethical ambivalences (as analyzed earlier). This will help to avoid under-complex, one-sided considerations and manipulative didactic approaches. Instead, students should be enabled to arrive at their own conclusions and make their own decisions for action based on knowledge and ethical considerations.

Of course, these considerations must be adapted to the respective age group. Thus, at the upper secondary school level, all of the facts and ambivalences analyzed earlier can be reflected on with pupils. However, a good way of dealing with ambivalent topics should already be initiated in primary schools, and so the didactic approaches chosen in this setting should make it clear to the children that the production and use of bioplastics is neither exclusively advantageous nor exclusively disadvantageous.

Furthermore, when dealing with the topic of bioplastics, the already mentioned problem regarding the individualization of responsibility for sustainable action should be counteracted. This means that pupils should not be given the impression that as individuals they are responsible for solving the problem of plastic waste, for which no final solutions have yet been found, even in the political sphere. It is important to empower pupils to make their own decisions and to show them ways to engage in sustainable consumption. At the same time, however, it must be made clear to them which actors from politics, economy, and society are in a position—and thus responsible—for setting the course at a higher level.

Since dealing with ambivalent issues in a promising way is quite challenging for teachers, good educational materials play a significant role. Following the abovementioned emancipatory approaches to education for sustainability in the sense of transformative learning, we understand educational materials as being pivotal for transparent transformation policies. We assume that transparent information in the context of socio-ecological transformation policies is crucial in enabling consumers and citizens to critically weigh up the advantages and disadvantages of bioplastics for themselves. But how helpful are existing educational materials on the topic of bioplastics? An exploratory case study presented in the next section provides initial answers to this question.

Explorative Case Study: Analysis of Teaching Materials for Primary Schools on the Topic of Bioplastics

In this section, we focus on a teaching material for primary schools on the topic of bioplastics entitled "Bioplastik selbst gemacht!" ("Bioplastics made by you!"), provided by the *German Federal Ministry for the Environment, Nature Conservation, Nuclear Safety and Consumer Protection* (*BMUV*). This ministry, one task of which is to inform consumers, provides free and adaptable educational materials (Open Educational Resources) on current environmental and sustainability-related topics that fit the curricula of different school subjects. These materials can be found on the website "Umwelt im Unterricht" ("Environment in the Classroom") (www.umwelt-im-unterricht.de).

On the occasion of the Science Year 2020/21, an initiative of the *German Federal Ministry of Education and Research (BMBF)* dedicated to the bioeconomy, this website made the bioeconomy its "Topic of the Month" in January 2020: "Bioeconomy—Sustainable Alternative to the Fossil Economy?" [translated by the authors]. Therefore, extensive teaching material on the subject has been published.[38]

The educational offer initially provides background information for teachers. First, a "Didactic Commentary" deals with the central characteristics and goals of a sustainable bioeconomy as well as the overall objective of dealing with the topic in class. Second, teachers receive information on the topic under the heading "Background" ("*Hintergrund*") about the relevance, potential, and risks of a bioeconomy, the role of science therein, requirements for a sustainable bioeconomy and the role of politics and consumers in implementing the concept. This is followed by two concrete teaching proposals, each consisting of instructions for teachers on how to conduct the lesson, as well as materials and tasks for the pupils. The first one is the already mentioned lesson plan for primary schools on "Bioplastics made by you!" and the second is a lesson plan for secondary schools entitled "Bioeconomy—New ideas for a sustainable economy" [translated by the authors]. In our analysis, we only focus on the primary school material ("Bioplastics made by you!"), as it explicitly refers to bioplastics and provides the background information for teachers mentioned earlier ("Didactic Commentary" and "Background").

Most primary schools in Germany comprise four school years and are attended by pupils aged six to ten. Environmental topics have been a central aspect of the German primary school curriculum for several decades, especially in the abovementioned "*Sachunterricht*," and in this context the ministry has developed materials on the topic of bioplastics. The fact that it is a material made available free of charge to teachers on a well-known website of the ministry responsible for the environment and consumer protection makes it a particularly interesting subject for our explorative case study. Our analysis aims to find out to what extent the ambivalences identified earlier are addressed and the material thus realizes an emancipatory approach to education for sustainability. The theoretical background of the analysis is the pedagogical theory of transformative learning[39] presented earlier.

The title of the theme of the month, namely "Bioeconomy—A Sustainable Alternative to the Fossil Economy?," is an open-ended question that implies that the aim of the educational approach is to weigh up the opportunities and limitations of a bioeconomy. The short introductory text under this heading then formulates it as a question to be addressed in class: "What opportunities does [the bioeconomy] offer—and why can it lead to conflicts?"[40] Accordingly, the "Didactic Commentary" also highlights the providers' claim that it wishes to familiarize students with the ambivalences of the topic. Thus, the overarching goal of both teaching concepts is to introduce students to the concept of a bioeconomy and to reflect on its potential as well as its limits and risks.

The subject-specific information provided for teachers ("Background") is based on a more precise definition and characterization of it as a normative concept ("The goal is a sustainable economy in which raw materials are used in closed material cycles"[41]). It not only deals with examples of existing bioeconomic products and the potential of such an economy (e.g., decoupling economic development from the consumption of finite fossil resources such as coal and oil; contributing to the reduction of greenhouse gases and a global transformation toward a sustainable economy) and the role of science (research on living organisms and development of new technologies), but also includes critical aspects. In this sense it is emphasized that "the concept is not uncontroversial, and critical voices on the side effects of a bioeconomy are increasing."[42] Several essential aspects, as summarized earlier, are explicitly addressed, albeit neither social aspects, such as unequal land access, nor the lack of participation in expert fora are mentioned.

However, despite these critical aspects, which are not explored in greater depth in the materials, the bioeconomy itself is not fundamentally called into question. Instead, the question formulated here is "Why is a sustainable bioeconomy important?," thereby making it a purely normative statement, i.e., a sustainable bioeconomy is important. The more fundamental question of whether the bioeconomy is a desirable path is not addressed.

The text emphasizes, however, that replacing fossil resources with bio-based resources is only one step toward sustainability and that a sustainable bioeconomy should have several characteristics. These include:

- Consideration of the limited availability of biomass as well as ecological and ethical criteria;
- Preservation of soil fertility and biodiversity, as well as no pollution of the environment, including water bodies;

- Priority of food security in the case of land competition;
- Extensive closure of energy and material flows in the sense of a circular economy;
- Reduction of resource consumption in all economic sectors (e.g., through changes in consumer behavior such as decreasing meat consumption, resource efficiency, and increasing yields on existing land).

However, all these aspects remain undefined, and the complex subject-specific interrelationships between them are only vaguely addressed. Moreover, the criteria mentioned are not discussed in terms of their feasibility. For example, the question of how to achieve a just and sustainable bioeconomy without degrading the environment and marginalizing poor local populations remains unanswered.

The bioeconomy concept is also not fundamentally questioned or discussed in terms of its potential for consumers. Instead, the text contains the clear statement "By providing bio-based alternatives to petroleum-based products, the bioeconomy fundamentally offers the possibility for more sustainable consumption."[43] How this statement is compatible with the critical aspects mentioned earlier remains unclear.

Concrete possibilities for the consumption of bioeconomic products are presented (clothing, sports equipment, bags, shoes, or children's toys made from bio-based materials such as pineapple leather or spider silk; food such as crisps made from insects; sausages made from algae; furniture made from hemp), but no more fundamental approaches to sufficiency or renunciation are included. The comment that bioeconomic products have not yet been uniformly certified and labeled (which leaves it unclear what consumers can orient themselves by) is followed by the extremely vague concluding note "Consumers can, however, make an effort to look into the respective product and their own consumption behaviour in order to assess the possible contribution to more sustainability."[44] The role and responsibility of consumers is not explicitly linked to the responsibility of policies or companies. Regarding the role of politics, the text only mentions German and European programs set up to promote the bioeconomy.

In order to assess more precisely to what extent this background information enables teachers to do justice to the topic of the bioeconomy (including its ambivalences) in the classroom in the sense of transformative learning, the teaching concept for primary schools on the topic "Bioplastics made by you!" is now analyzed.

In the introductory phase of the lesson, the teacher presents the following guiding questions to the students: "What is the difference between plastics and bioplastics? How can biomaterials help to produce environmentally-friendly products?"[45] There are no instructions on how to explain the terms "bioplastics" and "biomaterials" to the students, who may not be familiar with them. The second guiding question also introduces a normative framework: biomaterials can help to produce environmentally friendly products—without any more fundamental questioning or problematizing whether this is actually the case.

Furthermore, the pupils read the comic strip "Farmer Hubert and the Maize Bags"[46] in this first phase of the lesson (see Figure 12.1). With the help of this

Figure 12.1 Comic strip “Farmer Hubert and the Maize Bags.”

Source: ©FNR/Michael Ryba.

comic, they learn the difference between bags made from maize (bioplastic) and conventional plastic bags. Two children in the comic strip meet Farmer Hubert, who informs them that the bags in which he sells his popcorn are not made from plastic but from maize. To the children’s delighted reaction (“Boah! Cool!”), he explains that the great thing is that the bags do not end up on the rubbish tip but that he uses them to produce heat and electricity in his biogas plant.

The following pupil insights are expected here: "Bags made from bioplastics are made from plants" and "They can be reused for other purposes after use."[47] Based on this example, the teacher explains the concept behind the bioeconomy. Ambivalences that characterize the topic of bioplastics (or even biogas) are not addressed in this material; instead, all actors in the comic express their enthusiasm about the plastic bags made from maize. In terms of content, only the production and further use of bioplastics are addressed (but not the problems that go along with them), while energy aspects such as high inputs for production and recycling, or problems related to resource and land consumption, are not taken up.

In the second phase of the lesson, the pupils produce bioplastics themselves from starch, vinegar, water, and glycerin. The objective here is formulated very vaguely: "The students produce bioplastics in an experiment and thus deal with the term 'bioeconomy'."[48]

In the third phase, the pupils discuss the process behind producing bioplastics and their properties. They then work on the worksheet "Bioplastics and Co.: Products made from biomaterials" by ticking everyday objects that can be made from biomaterials. The aim of this phase is formulated as follows: "The material is used to introduce pupils to the possible variety of biological raw materials. Through the presentation of different products, the distinction between a biological and a fossil economy should be internalized."[49] It becomes clear that up to this point, no possible disadvantages of bioplastics have been addressed in class.

Students should discuss the advantages and disadvantages of such a production method only at the end of the lesson. For this purpose, the teacher should give the following food for thought:

> What is special about bioplastics compared to conventional plastics? Where do the materials for conventional plastic come from? What are the consequences of drilling for the oil for plastic? Where do the materials for bioplastics come from? What are the possible consequences of using foodstuffs such as maize in large quantities for bioplastics? What happens to plastic or bioplastics after use (waste)? What possibilities are there to reduce the above-mentioned consequences? What role does your own consumption play in this?[50]

At the end of the lesson, the students should then summarize "which option they find best to reduce the consequences of the production of not only bioplastics, but also plastics."[51]

All these questions demonstrate that some forms of topic ambivalences are implicitly addressed. However, it remains completely unclear on what basis the students are supposed to arrive at answers in this case. There is no professional input, no research assignment and no teaching material from which students can gain the necessary subject-specific insights, so it can hardly be expected that the not-so-simple controversies surrounding the topic can be adequately elaborated and understood in this way. Another problem at this point is that the subject-specific background information for teachers does not provide comprehensive material on the discussion points addressed here, so it remains questionable how the teacher can adequately lead such a discussion. Moreover, the proposed impulse questions do not explicitly aim at a real discussion but rather at clarifying factual questions.

Overall, the teaching proposal hardly depicts the ambivalences of the bioeconomy topic by using the example of bioplastics. The analysis reveals the following problem areas in the educational offering: students are not provided with any material for the targeted elaboration of relevant ambivalences adapted to their age group. As such this elaboration has to take place exclusively based on verbal ideas presented by the teacher. However, this seems barely feasible without any professional input, teaching materials, or research assignments.

The background information for teachers also deals only very superficially with the ambivalences. They are mentioned, certainly, but not explored in greater depth in relation to their contexts. Thus, it remains questionable as to what extent teachers can adequately support their students' learning processes without (time-consuming) further research. Sufficiency and renunciation strategies, i.e., alternatives to a consumption or technology orientation, such as "plastic fasting,"[52] are hardly brought into view or concretized.

Certainly, children of primary school age must not be overtaxed cognitively. At the same time, however, teachers are expected to discuss advantages and disadvantages with the children according to the objectives of the proposed teaching concept. This seems to be very important in view of existing ambivalences and is possible by means of an appropriate didactic reduction to central, comprehensible, and illustrative aspects. However, this educational offer does not provide enough support in terms of how such an age-appropriate didactic reduction might look, or how the prerequisites for such a discussion can be created.

Overall, the educational material thus only partially corresponds to current approaches to transformative learning. Such an understanding of education can be realized in a child-friendly way, even in primary schools. As this is an extremely demanding task, it requires all the more targeted support for teachers in the form of helpful background information and attractive teaching concepts.

Conclusion

As illustrated in this chapter, the example of bioplastics reveals the ambivalences of bioeconomic policies. On the one hand, the development of sustainable alternatives to conventional plastics is a necessary starting point for socio-ecological transformation toward a bioeconomy. On the other hand, bioplastics can exacerbate the existing environmental and social problems of plastic waste.

These ambivalences cannot be resolved easily—either technically or politically. Thus, there can be no one-size-fits-all solution; rather, we need a public discussion around the question of how we can reduce plastics in general. In this sense, the ethical perspective gives helpful directions for defining the guidelines of these societal negotiation processes. A norm-ethical approach that, for example, goes beyond banning plastic straws is no longer helpful than a virtue-ethical approach that imposes too much on the individual. Both have to be linked, as argued herein.

The precondition for an ethical decision on the individual and structural level is that all citizens are well-informed about the ambivalences of bioplastics. As argued earlier, transparent information, education, and empowerment start at primary school.

Focusing on the goal of enabling individual reflection on the ambivalences of bioplastics, the primary school is one of the most important places to enable transformative learning, which is the condition for handling these ambivalences. Transformative literacy, based on reflection and the deconstruction of prevailing power relations and narratives, allows one to stand up for their convictions and realize them through participation in democratic negotiation processes. This transformation in the mind can only succeed with the help of the presentation of ambivalences, as established in the case of bioplastics.

Instead of presenting bioplastics as the only solution to the plastic crisis, the following questions should rather be discussed with pupils as well as in society as a whole: how can we as individuals and society save plastics in general? Only against this backdrop can we then discuss the following technical question: which kind of bioplastics can be both useful and sustainable? Finally, it is important to note that the solution cannot come from the individual alone but that structural measures as well as clear guidelines are needed in order to address the plastic problem.

Notes

1 Pia Skoczinski et al., *Bio-based Building Blocks and Polymers – Global Capacities, Production and Trends 2020–2025* (Hürth: Nova Institute, 2021), 117; Check also "Plastic Facts & Figures," *Plastic Soup Foundation*, accessed October 10, 2023, www.plasticsoupfoundation.org/en/plastic-facts-and-figures/ as well as "Plastics – the Facts 2020," *Plastics Europe*, accessed September 5, 2023, https://plasticseurope.org/de/knowledge-hub/plastics-the-facts-2020/.

2 Steven Feit and Carroll Muffett, "Plastik heizt das Klima an," in *Plastikatlas 2019: Daten und Fakten über eine Welt voller Kunststoff*, ed. Heinrich-Böll-Stiftung and BUND (Berlin, 2021), 26f., www.boell.de/sites/default/files/2022-01/Boell_Plastikatlas%202019%206.Auflage_V01_kommentierbar.pdf.

3 Heather A. Leslie et al., "Discovery and Quantification of Plastic Particle Pollution in Human Blood," *Environment International* 163 (2022): 107199, https://doi.org/10.1016/j.envint.2022.107199.

4 Heinrich-Böll-Stiftung and BUND (eds.), *Plastikatlas 2019: Daten und Fakten über eine Welt voller Kunststoff* (Berlin, 2021), www.boell.de/sites/default/files/2022-01/Boell_Plastikatlas%202019%206.Auflage_V01_kommentierbar.pdf.

5 Research on plastics has been going on for 200 years, on bioplastics since the 1960s and the first company was founded in the 1990s. See "The History of Bioplastics," *Bioplastics News*, accessed August 4, 2023, https://bioplasticsnews.com/2018/07/05/history-of-bioplastics/.

6 "What are bioplastics?" *European bioplastics*, accessed April 19, 2023, www.european-bioplastics.org/bioplastics/.

7 BMBF – Bundesministerium für Bildung und Forschung and BMEL – Bundesministerium für Ernährung und Landwirtschaft (eds.), *National Bioeconomy Strategy* (Berlin, 2020), www.bmbf.de/upload_filestore/pub/BMBF_Nationale_Biooekonomiestrategie_Langfassung_eng.pdf.
8 Jan-Georg Rosenboom, Robert Langer, and Giovanni Traverso, "Bioplastics for a Circular Economy," *Nature Reviews. Materials* 7, no. 2 (2022): 117–137, https://doi.org/10.1038/s41578-021-00407-8.
9 Cf. "Bioplastics market development update 2023," *European bioplastics*, accessed April 19, 2023, www.european-bioplastics.org/market/.
10 Rosenboom, Langer, and Traverso, "Bioplastics."
11 OECD—Organisation for Economic Co-operation and Development (eds.), *The Bioeconomy to 2030: Designing a Policy Agenda* (Paris: OECD Publishing, 2009), https://doi.org/10.1787/9789264056886-en.
12 BMBF—Bundesministerium für Bildung und Forschung and BMEL—Bundesministerium für Ernährung und Landwirtschaft (eds.), *Nationale Forschungsstrategie BioÖkonomie 2030. Unser Weg zu einer bio-basierten Wirtschaft* (Berlin, 2010), https://biooekonomie.de/sites/default/files/files/2016-09/nfsb_2030.pdf.
13 BMBF and BMEL, *National Bioeconomy Strategy*.
14 Maria Backhouse, "Global Inequalities and Extractive Knowledge Production in the Bioeconomy," in *Bioeconomy and Global Inequalities: Socio-Ecological Perspectives on Biomass Sourcing and Production*, ed. Maria Backhouse et al. (Cham: Palgrave Macmillan, 2021), 25–44.
15 Maria Backhouse et al. (eds.), *Bioeconomy and Global Inequalities: Socio-Ecological Perspectives on Biomass Sourcing and Production (*Cham: Palgrave Macmillan, 2021), 4–5.
16 Kean Birch, Les Levidow, and Theo Papaioannou, "Sustainable Capital? The Neoliberalization of Nature and Knowledge in the European 'Knowledge-Based Bio-Economy'," *Sustainability* 2, no. 9 (2010): 2898–2918, https://doi.org/10.3390/su2092898.
17 Backhouse et al., *Bioeconomy and Global Inequalities*.
18 Ulrich Brand, Christoph Görg, and Markus Wissen, "Overcoming Neoliberal Globalization: Social-Ecological Transformation from a Polanyian Perspective and Beyond," *Globalizations* 17, no. 1 (2020): 161–176, https://doi.org/10.1080/14747731.2019.1644708.
19 Mario Giampietro, "On the Circular Bioeconomy and Decoupling: Implications for Sustainable Growth," *Ecological Economics* 162 (2019): 143–145, 143–156, https://doi.org/10.1016/J.ECOLECON.2019.05.001.
20 Hariati Sinaga, "Buruh Siluman: The Making and Maintaining of Cheap and Disciplined Labour on Oil Palm Plantations in Indonesia," in *Bioeconomy and Global Inequalities: Socio-Ecological Perspectives on Biomass Sourcing and Production*, ed. Maria Backhouse et al. (Cham: Palgrave Macmillan, 2021),175–192.
21 Backhouse et al., *Bioeconomy and Global Inequalities*.
22 Anne Tittor, "Towards an Extractivist Bioeconomy? The Risk of Deepening Agrarian Extractivism When Promoting Bioeconomy in Argentina," in *Bioeconomy and Global Inequalities: Socio-Ecological Perspectives on Biomass Sourcing and Production*, ed. Maria Backhouse et al. (Cham: Palgrave Macmillan, 2021), 309–330.
23 IPCC—International Panel on Climate Change (eds.), *IPCC Special Report on Climate Change, Desertification, Land Degradation, Sustainable Land Management, Food Security, and Greenhouse gas fluxes in Terrestrial Ecosystems. Summary for*

Policymakers Approved Draft (2019), www.ipcc.ch/site/assets/uploads/2019/08/Edited-SPM_Approved_Microsite_FINAL.pdf.

24 Christoph Lauwigi, "Mais statt Öl ist keine Lösung," in *Plastikatlas 2019: Daten und Fakten über eine Welt voller Kunststoff*, ed. Heinrich-Böll-Stiftung and BUND (Berlin, 2021), 35f., www.boell.de/sites/default/files/2022-01/Boell_Plastikatlas%202019%206.Auflage_V01_kommentierbar.pdf.

25 Venkateshwaran Venkatachalam et al., "Ökologische Nachhaltigkeitsbewertung von Biokunststoffen," in *Biokunststoffe unter dem Blickwinkel der Nachhaltigkeit und Kommunikation. Status quo, Möglichkeiten und Herausforderungen*, ed. Hans-Josef Endres et al. (Wiesbaden: Springer, 2020), 27–54, 48f., https://doi.org/10.1007/978-3-658-27795-6.

26 "Gemeinsam gegen Plastik," VerbraucherService Bayern im KDFB e.V., last modified June 23, 2016, www.verbraucherservice-bayern.de/themen/umwelt/biokunststoffe-nicht-automatisch-gut

27 Lisa Zimmermann et al., "Benchmarking the in Vitro Toxicity and Chemical Composition of Plastic Consumer Products," *Environmental Science & Technology* 53, no. 19 (2019): 11467–11477, https://doi.org/10.1021/acs.est.9b02293.

28 VerbraucherService Bayern im KDFB e.V., "Gemeinsam gegen Plastik."

29 Ibid.

30 Hannah Behnsen and Hans-Josef Endres, "Biokunststoffe – Hintergründe," in *Biokunststoffe unter dem Blickwinkel der Nachhaltigkeit und Kommunikation. Status quo, Möglichkeiten und Herausforderungen*, ed. Hans-Josef Endres et al. (Wiesbaden: Springer, 2020), 7–16.

31 Julia Blesin and Wiebke Möhring, "Bestandsaufnahme der öffentlichen Kommunikation und Wahrnehmung," in *Biokunststoffe unter dem Blickwinkel der Nachhaltigkeit und Kommunikation. Status quo, Möglichkeiten und Herausforderungen*, ed. Hans-Josef Endres et al. (Wiesbaden: Springer, 2020), 97–126.

32 Venkatachalam et al., "Ökologische Nachhaltigkeitsbewertung von Biokunststoffen," 48f.

33 Julia Petersen and Kerstin Schlögl-Flierl, "Plastikfasten – moraltheologische und sozialethische Anstöße," *Theologie der Gegenwart* 63, no. 2 (2020): 130–141.

34 Fabian Pettig and Ulrike Ohl, "Transformatives Lernen für einen sozial-ökologischen Wandel," *Praxis Geographie* 53, no. 1 (2023): 4–9.

35 Pettig and Ohl, "Transformatives Lernen."

36 Ibid.

37 Fabian Pettig, "Transformative Lernangebote Kritisch-Reflexiv Gestalten. Fachdidaktische Orientierungen Einer Emanzipatorischen BNE," *GW-Unterricht* 162, no. 2 (2021): 5–17 (6).

38 "Bioökonomie – Nachhaltige Alternative zur fossilen Wirtschaft?" *BMUV – Bundesministerium für Umwelt, Naturschutz, nukleare Sicherheit und Verbraucherschutz*, last modified January 30, 2020, www.umwelt-im-unterricht.de/wochenthemen/biooekonomie-nachhaltige-alternative-zur-fossilen-wirtschaft.

39 Pettig and Ohl, "Transformatives Lernen."

40 BMUV, "Bioökonomie."

41 "Hintergrund: Bioökonomie – Wirtschaften mit biobasierten Rohstoffen," BMUV – Bundesministerium für Umwelt, Naturschutz, nukleare Sicherheit und Verbraucherschutz, last modified January 30, 2020, www.umwelt-im-unterricht.de/hintergrund/biooekonomie-wirtschaften-mit-biobasierten-rohstoffen.

42 BMUV, "Hintergrund."

43 Ibid.

44 Ibid.

45 "Unterrichtsvorschlag: Bioplastik selbst gemacht!" BMUV – Bundesministerium für Umwelt, Naturschutz, nukleare Sicherheit und Verbraucherschutz, last modified January 30, 2020, www.umwelt-im-unterricht.de/unterrichtsvorschlaege/bioplastik-selbst-gemacht.

46 Michael Ryba, "Bauer Hubert und die Maistüten," Fachagentur Nachwachsende Rohstoffe e.V., accessed March 20, 2023, www.bauerhubert.de/index.php?id=9469.

47 BMUV, "Unterrichtsvorschlag."

48 Ibid.

49 "Bioplastik und Co.: Produkte der Bioökonomie: Arbeitsmaterial (Grundschule)," BMUV – Bundesministerium für Umwelt, Naturschutz, nukleare Sicherheit und Verbraucherschutz, last modified January 30, 2020, www.umwelt-im-unterricht.de/medien/dateien/bioplastik-und-co-produkte-der-biooekonomie.

50 BMUV, "Unterrichtsvorschlag."

51 Ibid.

52 Petersen and Schlögl-Flierl, "Plastikfasten."

13 Against All Odds

Managing Ambivalence in Philippe Squarzoni's Graphic Novel *Climate Changed*

Kirsten Twelbeck

Introduction: Creating Pockets of Possibility and Hope

Climate Changed is the English title of a 2014 graphic novel by Philippe Squarzoni,[1] a prize-winning French author of various politically daring "bandes dessinées." The 470-page book combines the voices of leading French climatologists, economists, and other experts with the thoughts and feelings of the narrator-protagonist—a graphic novelist who looks like the author, Squarzoni, and is working on a graphic novel about global warming. The overlap between narrator, protagonist, and graphic novelist evokes maximum credibility and intimacy and keeps together an otherwise fragmented, heavily disrupted narrative where expert talk alternates with mass-mediated scenarios of future social-environmental apocalypse, personal fantasies of (armed) environmental terrorism, and memories of a childhood in the beautiful Ardèche region (France). Stream-of-consciousness confessions, daydreams, and nightmarish projections of a dystopian future suggest a single persona searching for answers and solutions.[2]

These creative interventions are part of an agenda maintaining that science and politics alone cannot end global warming. The book embraces scientific data but it also seeks new, emotionally appealing strategies to inspire intellectual, emotional, and behavioral transformation. While it recognizes its own yearning for a once-and-for-all socio-ecological transformation, *Climate Changed* also acknowledges ambivalence as an emotional and cognitive constant among the human species, drawing attention to the many choices we make—and always will make—as individuals and as socio-political collectives in a world marked by permanent and accelerating socio-environmental crises.

More recent developments like electric cars, compensation for travel carbon emissions, or the marketing of more ecologically sustainable goods have not abolished the contradictions that the 2014 book[3] seeks to explore and solve. It chooses a very personal approach to the problem: readers witness the non-linear workings of a creative mind that not only seeks to understand climate research, but also observes and documents its own emotional ambivalence and contradictory behavior during this *Personal Journey Through the Science* (subtitle). Understanding human-made climate change in detail, knowing what has to be done

DOI: 10.4324/9781032627984-16

to keep it from getting out of control, and the wish to offset his CO_2 footprint do not keep the narrator-protagonist from driving a car, flying long distance and spending money on consumer goods. Importantly, however, the book does not blame the individual in the first place but seeks to integrate cognitive dissonance as a constant in a long-term mental, cultural, political, and economic transformation process. This chapter asks how the aesthetic means of the graphic novel make this mental integration conceivable.

For a general understanding of my analysis, knowing that *Climate Changed* follows (at least) two intertwined narrative strands is helpful. First, the book addresses narrative collapse in the context of global warming, and second, it traces the aesthetic-intellectual efforts of a visual artist-activist in disseminating knowledge, mental resilience, and hope in a world that is always and already (in) formed by the powerful cultural imagery of the West. How *Climate Changed* disrupts the mass-mediated status quo, how it brings in ambivalence and environmental justice concerns, and how it pinpoints the dilemmas of climate change communication will be the central pillars in this chapter.[4] Significantly, the narrator-protagonist prefers to ignore the everyday marginalization of victimized communities, in that salvation and action movies and advertisements have a soothing effect on his everyday thinking. As Ann Gardiner reminds us, this self-reflective approach enables readers to "understand the representational problems in narrating the story of climate change."[5] Importantly, however, *Climate Changed* seeks to create *new* meaning from such insights. By integrating paradox and ambivalence as integral aspects of human thought, emotions, and behavior, it not only challenges the normative, either-or approach to transformative action, but also creates new pathways to personal and collective agency. How these pathways appear, and what strategies the book relies on to carve out "pockets of hope and possibility" from within a context marked by ambivalence and contradictions, is the central research question of this chapter. As I demonstrate, *Climate Changed* conceptualizes the subject as socially produced and containing multiple, at times contradictory and competing subjectivities that unfold in response to climate change. To represent ambivalent ways of being-in-the-world, the book emphasizes the clashing of emotional and ethical choices and the simultaneity of seemingly incompatible emotions and events unfolding on a local and global scale. Its creative, self-reflective comments on the power of modern media culture and the way visuality informs the human mind are key to the book's creative analysis and problem-solving strategies. I therefore continue with some remarks about the role of seeing in *Climate Changed* to then show how Squarzoni challenges the narrative of development and progress by emphasizing simultaneity and ambivalence as drivers of modern cultures. In a third step, I discuss the book's representation of climate science as a seemingly stabile backdrop to an otherwise messy and ambiguous reality. The chapter ends with a close analysis of a sequence from *Climate Changed* to show how it "manages ambivalence" as a necessary component in the socio-ecological transformation process that is so urgently needed.

Defragmenting the Cluttered Mind

Climate Changed represents modern life as a mix of "past emotional-spatial experiences"[6] that it conveys via snippets from popular culture, (car) commercials, media footage, and scientific modes of representation (statistics, temperature curves, maps, satellite images). As the media and communication scholar Marita Sturken reminds us, all of these images permeate our mass-mediated lives in the West, thus reinforcing ideological constructions of nature, human relationships, and culture.[7] *Climate Changed* challenges these established "practices of looking": by rearranging, recontextualizing, and overwriting familiar imagery, it discerns the cultural and economic drivers of human-made global warming, thereby complicating the seemingly paradoxical relationship between scientific knowledge and political (in)action.

This "defragmentation" is inseparable from the book's critique of Western media platforms displaying climate change as external to Euro-American lives. By recontextualizing drawn copies of media footage about drought refugees in the Global South and Hurricane Katrina, and even integrating a photo of collapsed sanitation systems (255), *Climate Changed* pushes readers to reflect on clichéd representations of economically and ecologically disadvantaged and politically marginalized communities abroad, as well as within countries of the "developed West." In what are essentially two variations on one theme, the book displays the silhouettes of Western men and women standing on dried-up land or (in the second panel, or individual frame) knee-deep in water, surrounded by media images of environmental and social collapse (251 and 294). The caption to these panels, each of which covers an entire page, reads "AND WE DON'T NOTICE A THING." By representing what remains unsaid (or is silenced), and by making visible what remains invisible (or is covered up) by the internalized images and emotions that flood the human brain, *Climate Changed* visualizes the paradoxical attitude that marks the Western being-in-the-world, i.e., the emotional and cognitive dissonances that precede, and color, the decisions we (and I include myself here) make.

The representational options that result from image-text combinations augment the complexity of the narrative. Some panels come without text and rely entirely on *showing* (reminiscent of a camera), while others introduce an invisible yet omniscient narrator who *explains* what the main protagonist is thinking.[8] Facilitated by the overlapping of narrative perspectives, *Climate Changed* switches with ease between the subjective views of the narrator, who tells his own story (the so-called autodiegetic narrator), and narration reminiscent of a voice-over in film (so-called heterodiegetic narration[9]) that often occurs as comment. Scientific facts and individual feelings are not only presented side-by-side, but also commented upon with the help of strategies that are particular to the graphic novel as a genre. It is in the tension between text and image, as well as in the gaps between individual frames (experts refer to the graphic novel's spatial organization as "gutter"), that ambivalences, contradictions, and paradoxes can freely unfold.

Between Anticipation, Forgetting, and Nostalgia: Grounds for Hope and Action

A vast panoply of non-chronological, time-related imagery and text further augments the book's emphasis on the incommensurate and contradictory. The linear trajectory of future climate trends is combined with feedback loops excavating nostalgic dreams, childhood memories, sedimented imagery from popular culture, as well as with conversations, walks, and vacations that were part of Squarzoni's life while working on his book. By inserting such imagery into sequences that document the graphic novelist's conversations with climate scientists, or that show the daily routines of the artist at work, the book creates intermediary, largely undefined spaces "between anticipating and forgetting the future" that are marked by uncertainty.[10] By visualizing associations that nest almost unnoticed in the human brain, even during phases of mental concentration, and by introducing cultural imagery that transforms scientific information into catastrophic thinking, *Climate Changed* creates imaginary sites of inspiration without denying the severity of the status quo. As the cultural critic and writer Rebecca Solnit puts it, such in-between and thoroughly ambivalent spaces provide imaginary "grounds for hope and action."[11]

Creating such spaces is crucially important for a book seeking to look anthropogenic climate change in the eye without furthering the frustration and inaction that are typical symptoms of "solastalgia," a form of depression caused by environmental decline.[12] *Climate Changed* addresses solastalgia by displaying the despair and the "hedonism or apathy"[13] of its eco-guilty protagonist-narrator. By inserting childhood photographs and tabloid advertisements from the 1950s featuring happy road trips with family and friends (284–285), it shows how a feel-good atmosphere of (bygone) possibilities and (naïve) innocence can serve as a mental refuge. The late sociologist and philosopher Zygmunt Baumann identified such a "retrotopian" reaction as a central paradigm of our "globalized" era:

> It is now the future, whose time to be pillorized seems to have arrived after being first decried for its untrustworthiness and unmanageability, that is booked on the debit side. And it is now the past's turn to be booked on the side of credit—a credit deserved (whether genuinely or putatively) by a site of still-free choice and investment of still-undiscredited hope.[14]

Its profound ambivalences keep *Climate Changed* at a safe distance from Baumann's pessimistic diagnosis, in that the book simultaneously embraces and rejects the lure of that "ache" for an idealized past (see Figure 13.1). The novel is positioned on the intersection between "restorative" and "reflective" nostalgia. In Svetlana Boym's definition of the term, "restorative nostalgia" pretends to "rebuild the mythical place called home."[15] In *Climate Changed*, it appears in the staging of the protagonist's carefree boyhood spent among toy cars and tractors, or when his right hand is shown in closeup, touching the grass: "THIS FAMILIARITY… THESE FEELINGS… HERE. I PLAYED IN THIS TALL GRASS…" (24).

Figure 13.1 An instance of restorative nostalgia.

Source: From *Climate Changed* by Philippe Squarzoni. *Saison Brune* by Philippe Squarzoni © Éditions Delcourt. 2012 Text and illustrations by Philippe Squarzoni. Translation by Ivanka Hahnenberger.

Many pages later, the book stages the very same scene as an instance of reflective nostalgia. Defined as a reaction to the loss of a place through real-life circumstances, there is now a sense of defamiliarization[16] in the panel showing the narrator as a young boy. Through the composition between image and text, the "relationship between past, present, and future"[17] takes center stage: "WE KEEP GOING BACK IN TIME. REPEATEDLY." That the book finds restorative nostalgia inacceptable without its reflective sibling is made explicit in the adjacent panel that reads "AS IF TO ESCAPE THE MARCH OF TIME" (96), accompanied by the image of a clock.

Besides learning to both *enjoy and reflect* upon nostalgia simultaneously, the reader is, time and again, shocked into the awareness that the trap of present-day mobility and consumerism is a product of past ideals. Images that rely on remediation (the "representation of one medium in another medium"[18]) pop up unexpectedly, highlighting the critical side of the idealized 1950s and the revivalism of that era in the early 1970s, during Squarzoni's childhood, when advertisements celebrating US car culture and consumerism were omnipresent. The reader takes the perspective of Walter Benjamin's famous "Angel of History," namely, that allegorical critique of the modern myth of progress and linear time turns its face toward the past while being propelled "into the future to which his back is turned, while the pile of debris before him grows skyward." The storm that pushes the "Angel of History" away from what he perceives as "one single catastrophe which keeps piling wreckage" is "what we call progress."[19] By emphasizing the ambiguity that can emerge in encounters with things once held dear, the book suggests an accepting and yet critical take on the past: daydreaming about a bygone time of hope and innocence is a fact of life and invigorating, but dealing with it responsibly, by keeping an eye on the formative impact of that past on our future, is marked as crucially important.

The Emotional Messiness of Decision-Making

Juggling simultaneity and ambivalence is relatively common in an artform that thrives on creative tensions between image and text, between one panel and the next, and between sequences. What distinguishes *Climate Changed* from other graphic novels (including Squarzoni's previous works) is the sense of urgency that haunts conversations and negotiations concerning climate change, including this one: how long will we talk about climate change instead of making the changes that are necessary to prevent socio-ecological collapse? Instead of answering the question, *Climate Changed* complicates it, along with the concept of decision-making itself. Throughout the book, decisions are represented as temporary, inconsequential, and emotionally messy. They emerge as a constant in the life of the informed citizen from the Global North, who wavers between the desire to live a modern, mobile life and the awareness that such a life is excessively energy-intensive and threatens the lives and livelihoods of billions of people. The reader, who is bombarded with the scientific evidence of anthropogenic climate change that Squarzoni also provides, can easily identify with the dilemma on display.

Instead of proposing a rigorous regime of (self-)control to replace the inconsequential mood of the present, *Climate Changed* dwells on the schizophrenic nature of human thought and action, thereby stretching the time that keeps the reader in an agonizing state of indecision. The book stages the protagonist as wavering between eco-guilt and blaming multinational corporations, the media, and consumer culture. The main targets are the industrialized societies in the so-called Global North:

> [...] THE TYPICAL FRENCH PERSON PRODUCES ON AVERAGE 2,7 TONS OF CARBON EQUIVALENT A YEAR WITH ALL OF THE GREENHOUSE GASES COMBINED... 1,8 TONS OF IT CO_2 ALONE. [...] AND IN MOST AFRICAN COUNTRIES, IT WAS 0,1 TONS PER PERSON PER YEAR.
>
> (158)

The quote appears on three large "moment-to-moment" panels that decelerate and intensify the reading process.[20] This "slow-motion" effect adds to the irony that these thoughts are focalized from behind a car windshield that the reader associates with the protagonist's vehicle. The paradox is explained, then, by the fact that modern life creates multiple subjectivities: shortly after witnessing the somewhat pathetic behavior of the prototypical "climate sinner," the former driver is shown as a pedestrian in a state of angry consternation. The panel reads "EVERY TIME I SEE A 4×4 OR AN SUV, I GET ANNOYED" (182). His anger turns into sarcasm as he imagines other drivers uttering flimsy, stereotypical excuses: "ONE MORE ON THE ROAD ISN'T GOING TO MAKE A BIG DIFFERENCE" and "I DON'T SEE WHY I SHOULD BE THE ONE WHO DEPRIVES MYSELF" (182). Later, however, the protagonist's moral superiority melts into air: when he is invited to Laos, he wonders why he "SHOULD GIVE UP THIS TRIP; WHILE OUT THERE 3,000 SOME ODD PEOPLE ARE RUNNING AROUND THE CITY IN 4-WHEEL DRIVES!" (208).

To the reader's relief, he eventually declines the invitation. The book suggests a knowledge-driven moral act, and in the 20-odd pages that follow the previous quote, *Climate Changed* illustrates the consequences of industrial pollution and melting glaciers, the connection between modern consumerism and brutal labor conditions, and the unequal distribution of responsibilities on a global scale. Readers more than follow this familiar type of disaster reporting, as they participate in the thought process that causes the protagonist's decision to call off the residency. Importantly, however, they are denied a heroic solution, as the protagonist has already related that "TO AVOID CLIMATE CHANGE, THE INHABITANTS OF RICH COUNTRIES MUST IMMEDIATELY DECREASE THEIR EMISSIONS... TO THE LEVEL OF A MALNOURISHED PERSON. NO ONE WANTS THAT. I CERTAINLY DON'T" (194). How limited his willingness to personally sacrifice his Western, middle-class living standard actually is, becomes clear when immediately after his "heroic moment," the protagonist is shown, together with his wife, on a visit to New York City. As the narrator-protagonist explains, tongue-in-cheek, the couple wanted to see places from "THE FILMS WE SAW WHEN WE WERE YOUNG" (230).

Despite its ironic debunking of pseudo-reasonable personal excuses and a high level of self-reflexivity, *Climate Changed* maintains that "CLIMATE CHANGE IS NOT JUST A MATTER OF INDIVIDUAL BEHAVIOR" because "THE PROBLEM LIES IN HOW OUR WHOLE SOCIETY IS STRUCTURED" (184). It is important to note that individual decisions to reduce CO_2 emissions are not declared superfluous here; rather, the book performs a shift in focus: by representing decision-making as profoundly ambiguous, *Climate Changed* redirects the reader's attention to the feelings that steer the Western experience in the 21st century, and to the structures that keep in place these feelings and the way of life that goes along with them.

Climate Changed reminds readers of their own "SPLIT PERSONALITY. STUCK… BETWEEN TWO SEASONS… BETWEEN TWO CONTRADICTORY STORIES... THAT WE'RE TELLING US SIMULTANEOUSLY" (284). Instead of implementing a new, systematic order, the book maintains that "TO ESCAPE THE DELUSIONAL STATE THAT HAS TAKEN HOLD OF US, WE NEED TO BREAK OUT OF CERTAIN MODES OF THOUGHT" (285). To enable (rather than force) this process, it practices (rather than demands) a questioning and re-coding of established norms, beliefs, practices, and priorities. By rearranging text and imagery in ever-new combinations, *Climate Changed* destabilizes established meaning-making practices and experiments with new modes of storytelling that react more adequately, self-critically, and responsibly to the messages that are communicated via droughts, floods, and climate refugees. By highlighting the contradictions and ambivalences of individual and collective behaviors that are a part of modern life, the book constitutes a starting point for mental renewal: only by recognizing the serious flaws of modernity, along with the natural instability of its paralyzing absolutes, can decisions be revised and understood as opportunities to reduce CO_2 emissions.

Taking a Break When There's No Time to Lose

As has been shown, behavioral and emotional inconsistencies are regular companions to modern life; therefore, a yearning for order is not at all surprising. *Climate Changed* translates this interconnection into a structural device. By alternating chronological narration with sudden moments of indecision and ambivalence, it unlocks "niches of possibility" for mental renewal. Some of these "niches" remain radically unscripted, like the all-black panel that signposts as-yet-undefined risks and possibilities: "SOME MORNINGS… BETWEEN SLEEPING AND WAKING… THERE IS A MOMENT OF UNCERTAINTY" (357).

Climate Changed relies on such moments to remind readers that complex ecological and societal transformation processes are particularly vulnerable to unforeseeable developments. It even insists that uncertainty exists despite, or along with, scientific consensus (by stressing, for instance, that climate change manifests itself quite differently on the global and local level). Speeding up administrative processes to build a "green" energy infrastructure may result in political division and paralysis, while technological innovation can have unexpected or underestimated effects on society and environment. The way modern media platforms impact public opinion is also hard to assess. Since considering such unresolved challenges will invariably slow down decision-making processes, the book's stubborn insistence on uncertainty is significant indeed. It even quotes one of the authors of the fourth and fifth IPCC report, Hervé le Treut, dampening expectations toward modern science: "WE DON'T REALLY KNOW HOW CLOSE WE ARE TO BREAKING THE THRESHOLDS OF THE MAJOR CLIMATE MECHANICS [...] LOCAL VALUES ARE STILL UNCERTAIN" (113). By including the incorruptible, self-critical voice of a prominent climate scientist, Squarzoni opens a tiny yet important "niche of possibility" to the exhausted reader.

By creating such niches, *Climate Changed* offers an alternative to the cultural obsession with reducing moments of uncertainty. The protagonist is a prototypical example of this obsession, in that throughout the book, he struggles—and fails—to overcome his own ambivalent thoughts, feelings, and behaviors. Zygmunt Bauman identified this struggle as the main feature of a "modern consciousness." In its extreme form, it becomes an obsessive "Quest for Order" marked by "restless action"[21] and

> the suspicion or awareness of the inconclusiveness of extant order; a consciousness prompted and moved by the premonition of inadequacy, nay non-viability, of the order-designing, ambivalence-eliminating project; of the randomness of the world and contingency of identities that constitute it. Consciousness is modern in as far as it reveals ever new layers of chaos underneath the lid of power-assisted order. Modern consciousness criticizes, warns, and alerts. It makes the action unstoppable by ever anew unmasking its ineffectiveness. It perpetuates the ordering practice by disqualifying its achievements and laying bare its defeats.[22]

The "modern consciousness" should not be confused with "a fight of one definition against another, of one way of articulating reality against a competitive proposal," because in modern cultures, "order" is "a fight of determination against ambiguity, of semantic precision against ambivalence, of transparency against obscurity, clarity against fuzziness."[23] Bauman, who was a Holocaust survivor, considered the Shoah the horrific culmination of this modern obsession with order. It is from this perspective that he considers ambivalence to be the "waste of modernity."[24]

Climate Changed welcomes the "waste" of the modern cultural text by embracing "polysemy, cognitive dissonance, polyvalent definitions, contingency; the overlapping meanings in the world of tidy classifications and filing cabinets."[25] Importantly, however, it equally acknowledges the desire for chronology, causal connections, and orientating principles as useful cultural constants. By choosing *A Personal Journey Through the Science* as the subtitle to the English version, it even inscribes itself in the literary tradition of narrative quest, a genre that traditionally combines individual growth and adventure. Ironically, however, *Climate Changed* merely appropriates the genre to subvert the dichotomy between order and ambivalence. By allowing their coexistence, it allows for hope and human agency to sit next to despair and failure: "SOMETIMES… IN THE MIDDLE… WE NEED TO TAKE A BREAK" (96).

"Taking a break" counteracts the sense of futility and hopelessness that tends to emerge because "THE CLOCK IS TICKING. AND TIME IS RUNNING OUT" (148). *Climate Changed* uses stereotypical jargon and irony to ward off climate change paralysis. "Empty" questions like "HOW TO BEGIN?" (148), "WHICH COURSE TO CHOOSE? HOW TO STICK TO A DECISION?" (242) serve a similar purpose. The book thus attacks the way contemporary debates are framed as being "pressed for time": why act at all if it is already "too late"? Furthermore, the book challenges the usefulness of such discourse and proposes de-temporalization instead. Readers are asked to reconsider their "MODES OF THOUGHT": why do we care so much about beginnings, endings, and finiteness, if climate change is ongoing? The book replaces time-related notions of continuity and progress through an "associative logic"[26] that privileges simultaneity, irreconcilable opinions, and contradictory behaviors over the illusion of neat and clear-cut chronological thought-processes. *Climate Changed* adds new beginnings without overwriting the previous one(s), and it ends repeatedly, only to resume the issue of transformation from a different angle. It includes memories of the past and future projections as part of the everyday workings of the mind and inserts remediated[27] film sequences that close with a finalizing "THE END" (302–303) before continuing the story outside the movie theater. Although it emphasizes ambivalence and inconsequential behavior, it maintains a (weak) notion of development, since it merely destabilizes its absoluteness and progress-orientation. By suggesting a nonlinear concept of development that relies as much on memory loops, daydreaming, and movie-viewing as on scientific models of foreseeable futures, *Climate Changed* responds to the paralyzing effects of modern consciousness by carving out imaginary sites of resilience from uncertainty.

Moving in and beyond "the Science"

Despite its emphasis on ambivalence, *Climate Changed* is firmly grounded in science. The book *documents* rather than *narrates* climate research, using a dry, matter-of-fact way of reporting. By doing so, it aligns itself with a scientific ethos: interviews are rendered in text-intensive speech balloons and via moment-to-moment sequencing (an expert turning his head, a closeup of her hands) as if to stamp scientific knowledge on the reader's mind. No detail diverts the recipient's attention from the experts' serious faces and the alarming information they share. These long and static sequences make *Climate Changed* a challenging—and at times exhausting—read.

Familiar (media) images of droughts and flooding, and abstract forms like graphs and temperature curves, do not ameliorate this readerly fatigue.[28] The only narrative feature that occasionally disturbs the hermetic modes of scientific thinking and environmental apocalypse are single frames or short sequences featuring fictional fantasies.[29] These insertions, or "little stories" (my terminology), visualize an emotional imbalance that includes ambivalent feelings (concerning, e.g., consumer culture or violent forms of environmental activism) that befall "ordinary people" who despair about the lack of meaningful political action in the face of climate catastrophe.

The "little story" that I wish to discuss is randomly inserted in an interview with the prominent French climate scientist Jean Jouzel and visualizes the mental karaoke that comes with catastrophic information about climate change (see Figure 13.2). A complex image with little text, the respective panel is a peculiar mix between realistic representation and fantasy. It features a smiling Santa Claus in the style of a comic-book figure, with rounded edges and soft features (368).

An icon of United States consumer culture, Santa Claus is lifting a huge white cloth, thereby dividing the panel into an upper and a lower part. His bearded face lurks from between skyscrapers and airplanes that fly toward the viewer in a quasi-military formation. Below the cloth, two people (easily identifiable as the protagonist and his wife) emerge, in ordinary streetwear, trying to block a car that is speeding toward them. The powerful technology makes them appear weak despite being armed. Like other space-consuming panels in this book, the image disrupts the lecturing mode of science communication and asks significant questions about consumer culture, the nature of mobility, and the futility of violent climate activism. Spectacular, disturbing, and seeped in sarcasm, it conjures up the aggressive atmosphere that comes with deregulated markets, political inaction, and sensationalist media reporting about the effects of climate change.

Importantly, however, *Climate Changed* self-reflectively stages the scene as straight out of an imaginary action movie whereby, under the neoliberal umbrella (cloth) of an endless shopping season, the citizen-as-action-hero replaces the politically capable state. The performative appearance of this "action panel" cautions against clichéd solutions to end global warming. At the same time, however, it provides an interim space in which to connect with extreme feelings—or rather affects[sic]—of fear, anger, and despair that tend to be neglected in climate change debates. Together with a few smaller, introductory panels, this eye-catching image enables readers to recognize climate change as *emotional* crisis. The effect of the insertion is stunning: when Jouzel resumes his speech, his earlier suggestion to "LOOK AT ACTUAL NEEDS" to "THEN ATTEMPT A RESPONSE OF MUCH LOWER CONSUMPTION WHILE AT THE SAME TIME DEVELOPING ENERGY SOURCES THAT ARE LESS RISKY AND CHEAPER" (367) no longer sounds abstract and formulaic. Under the influence of Hollywood apocalypse, "ENERGY-CONSERVATION POLICIES," helped by "SIGNIFICANT FINANCIAL INVESTMENT" (373), become not only intellectually comprehensible, but also emotionally desirable.

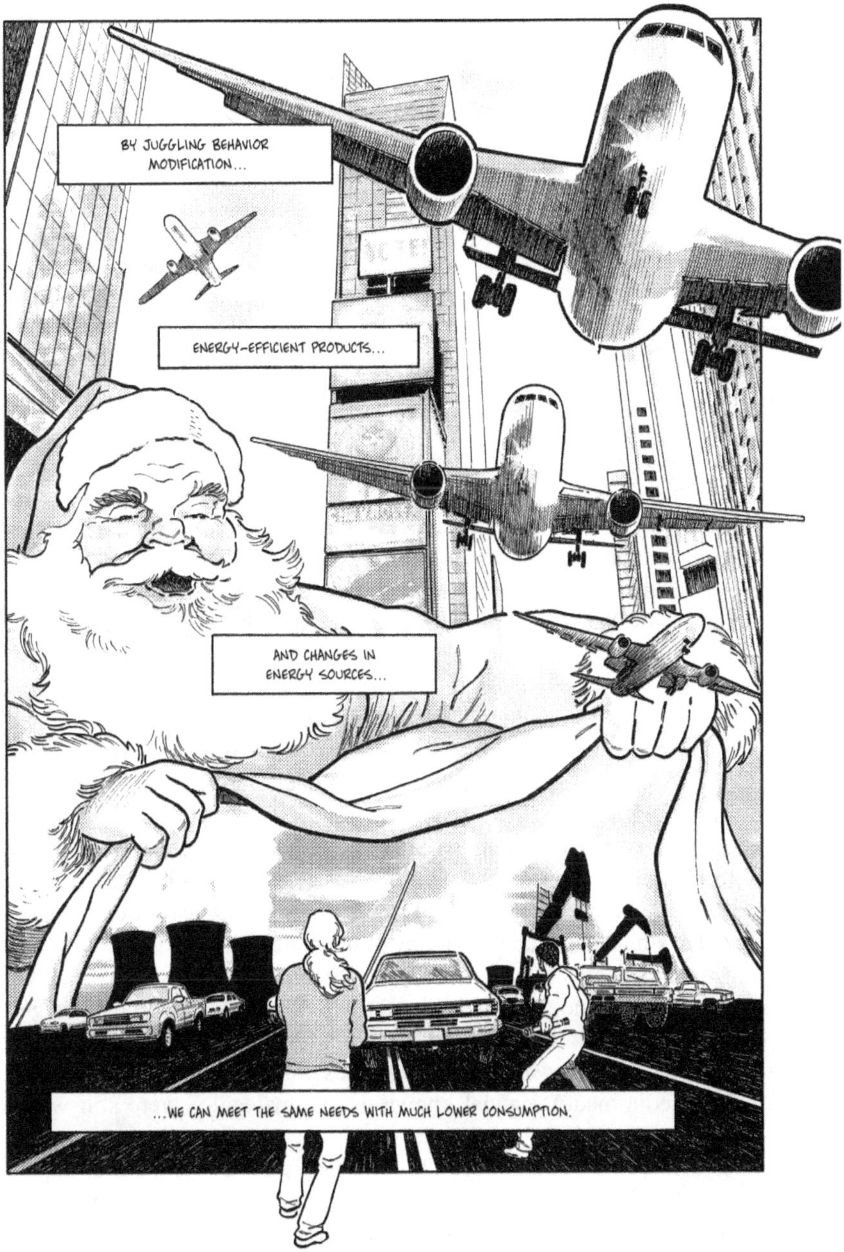

Figure 13.2 At war with Santa Claus.

Source: From *Climate Changed* by Philippe Squarzoni. *Saison Brune* by Philippe Squarzoni © Éditions Delcourt. 2012 Text and illustrations by Philippe Squarzoni. Translation by Ivanka Hahnenberger.

Despite its cautionary take on "eco-terrorism,"[30] *Climate Changed* does not defy radical forms of climate activism (see Figure 13.3); instead, it relies on ambivalence as a form of provocation. Tying in with Jennifer D. Grubbs' analysis of the "good and bad activism" dichotomy in public discourse, the graphic novel relies on subversive, "queer" ways to create the "stimulating images" of human rage and sensation that contemporary mass-mediated societies crave.[31]

As an analytical text, *Climate Changed* distinguishes between scientific findings and emotional reactions to these findings. Those who are (ideally) humble, clear, and unambiguous by profession are set apart from the "ordinary people" who tend to react emotionally, sometimes inconsistently, and in a contradictory manner in terms of their thinking and behavior. While this makes the "average citizen" appear more "human" compared to the figure of the "expert,"[32] the latter's problem-solving competencies are highly acclaimed throughout the book, along with the descriptive language that prevails in natural science communication.

How does this relate to the restructuring of established modes of thought? Moreover, how does this book, which cherishes uncertainty, deal with the fact that experts continuously seek to improve and correct previous projections? Are they not practitioners of that modern striving for order that leads straight into a world where there are no alternatives, no tolerance for contradictions and ambivalences? Let us bear in mind that Baumann does not condemn the striving for ordering knowledge systems as such but that he draws attention to their potential to destroy, and even to eliminate, ambivalence, in the service of an absolute "purity." In line with Bauman's warning impetus, *Climate Changed* celebrates scientific knowledge systems that are profoundly aware of the dark side of modernity. The book represents "the Science" as a flexible, international, multi-voice system that follows a fine-tuned ethos of responsibility. Readers "witness" a conversation between the renown French climatologist Hervé le Treut and the environmental journalist Hervé Kempf about the mechanisms of control that precede the publication of results in climate research. Le Treut embodies the ideal of the "responsible scientist" when he concedes that many of his colleagues fail to communicate their findings in understandable ways (102), and then patiently describes scientific processes, regulations, and rules of conduct (86–87). By explaining the genesis of peer reviews and calling to mind the availability of scientific debates and disagreements, this idealized figure reestablishes trust in science: in 2012, the effects of the (media-made) scandal known as "Climategate" were still widely felt.[33]

As the scientists in the book point out, there is hardly any government or population that fails to realize that human-made global warming *is* indeed a problem (88). What is lesser known is that in 2012 (the year when the French original, *Saison Brune*, was published[34]), older data had turned out to be surprisingly accurate. It is this awareness of having been ignored in the past that lets experts proclaim that "WE'RE ALREADY THERE," and that "THE CLIMATE FOR THE NEXT TWENTY YEARS IS ALREADY LAID OUT. WE WON'T BE ABLE TO AVOID

Figure 13.3 The author's wife as action heroine.

Source: From *Climate Changed* by Philippe Squarzoni. *Saison Brune* by Philippe Squarzoni © Éditions Delcourt. 2012 Text and illustrations by Philippe Squarzoni. Translation by Ivanka Hahnenberger.

IT" (106). This, then, is the time for the graphic novelist to showcase the creative means and strategies of his genre. In an age that is already way ahead of what is indicated on clocks and calendars, Squarzoni embraces the role of helpmate to inform, reflect, and offer solutions.

Managing Ambivalence

How exactly does this story, which is so saturated with socio-environmental decline and collapse, create a sense of hope or agency? As I show in a final, exemplary analysis of one sequence, *Climate Changed trains* readers to *manage* ambivalence by *normalizing* indecisions and helplessness and by *enhancing* the creative and innovative potential of the as-yet-undecided.

Normalization is facilitated by spatial organization, in that the graphic novel resolves the theoretical opposition between change and continuity by forcing upon the reader its structural breaks, pauses, and moments of indecision. The author-narrator-protagonist's mind starts wandering, gets back into focus, trails off, changes direction, and then gets back into action mode. Seemingly indifferent to his surroundings, the protagonist-narrator observes, listens, remembers, panics, makes behavioral choices, and anticipates political solutions. By representing the seemingly passive states of waiting and aimless wandering as a multidirectional activity of a creative mind, the book turns presumed weaknesses of the human species into a prerequisite for meaningful action.

To explain how this trains readers to become more susceptible to the creative dimensions of indirection and ambivalence, I have selected six evenly arranged panels on one page (97; see Figure 13.4). Although they appear in pairs (the pattern is known as a 3×2 grid in comics studies[35]), there is little connection between them. Readers are free to choose alternative viewing paths,[36] each one of which results in a different story.

One of those paths connects the three vertical panels[37] on the right because they all show human and animal faces: a child, a man, a dog (they can easily be identified as the young and adult author and his dog). This, then, is a story of personal evolution and cross-species connectivity. A more philosophical reading of those images diagnoses an almost unbearable, manifold tension between hope (the protagonist-as-a-child, smiling), responsibility (the protagonist-as-an-adult gazing toward the adjacent panel that shows the half-lit planet earth), and resignation (the family dog, that will, later in the novel, be found dead). According to such a fatalistic reading, realizing that we will most likely exceed 1.5 degrees Celsius of global warming[38] becomes a quasi-natural tipping point in a linear evolution toward death. Interestingly, however, the visual-textual complexity of the sequence complicates this all-too-familiar narrative of decline. According to the non-relational text boxes (they lack the tail of a speech balloon and function as a meta-commentary) that are randomly strewn across the six panels, we are in a perpetual state of in-betweenness where the end is yet to come:

Figure 13.4 Readers can choose between alternative viewing paths.

Source: From *Climate Changed* by Philippe Squarzoni. *Saison Brune* by Philippe Squarzoni © Éditions Delcourt. 2012 Text and illustrations by Philippe Squarzoni. Translation by Ivanka Hahnenberger. English translation copyright © 2014 Harry N. Abrams, Inc. All rights reserved.

> WE'RE IN BETWEEN TWO STORIES. THE FIRST WAS FULL OF HOPE. OF PROMISE. THE FIRST WAS WIDE OPEN. IT FACED TOWARD THE SECOND PART. THE SECOND; THE PART WHERE WE HAVE TO MAKE A COMMITMENT, IS UPON US. SOMETHING MENACING HAS TAKEN THE PLACE OF HOPE. AN ENDING THAT LIMITS OUR PROMISE. THERE'S NO WAY OUT OF IT. ALL THINGS COME TO AN END.
>
> (97)

The negativity that emanates from these comments breathes melancholia rather than doom and depression. There is, in fact, a certain lightness, and even a slight sense of humor, emerging from the sequence: the multidirectional relationship between the text snippets and imagery add a sense of playfulness and happy contemplation to the melancholic mood of these six images. The sequence challenges the absoluteness of apocalypse along with the idea that living "between two stories" must necessarily feel like an existential trap. The loose connection between the three portraits mentioned earlier relies on variation—a strategy that highlights the creative potential of repetition: a "frontal shot" of the young boy, the profile of his adult self, followed by a side view of the family dog. Due to the relative disconnect between these images, readers can easily exit from the *one* pessimistic story and delve into childhood nostalgia instead. They can bypass the fatalistic story mentioned earlier and decide to dwell on each individual panel instead, enjoy the landscape, admire the dog. They can also resort to a horizontal reading: the regular, 3×2 grid structure encourages three pairings of two panels to create a narrative link between child and landscape, or between the man's gaze and the clichéd picture of Planet Earth. The last pair connects the dog with the pitch-dark geometrical square in the lower-left corner containing the caption "AN ENDING THAT LIMITS OUR PROMISE." There is irony in this combination, since dogs are (apart from being conditioned to react joyfully when hearing someone open a package of dog treats) quite unaware of the promises that drive their humans.

Obviously, this sequence is telling several stories simultaneously, encouraging readers to choose between hopelessness, contemplation, and further storytelling. They can read the final words ("COME TO AN END") as either an announcement of apocalypse or as indicating a nightmare known as "global warming" that will stop as soon as the alarm goes off. They can build a narrative of nostalgic yearning or rely on the memories of their own hopeful childhood as a motor for socio-ecological transformation. They can even try to do away with endings, and beginnings, and focus on what *is*. The book alerts them to the normalcy of simultaneity and contradiction, and by creating open structures and multiple meanings, it inspires playful and creative practices of reading. In the end, it is the recipient's individual responsibility to manage ambivalence by creating a meaningful story and, one may add, a story to live by in an era of perpetual precariousness.

In Place of an Ending

According to a very early panel in the graphic novel, "THE HARDEST THING" will be "... HOW TO END" (14) the protagonist's "personal journey." About 460 pages later, he does not seem to have come far at all; in fact, he seems to remain the same person throughout the novel, namely, a sympathetic artist who cherishes the "JOY OF CONSUMING" (he loves to buy books, 37), owns a dog (high carbon footprint!) and a (large) fridge, and takes the car to Ardèche to create a graphic novel about global warming. This is where we see him, at the very end of the book, closing the lid of a coal-burning (!) stove before he resumes work on his (energy-intensive) computer. The following dialogue is the final text part in *Climate Changed*: [He:] "YOU KNOW THAT WOODY ALLEN QUOTE? I WISH I COULD THINK OF A POSITIVE POINT TO LEAVE YOU WITH. WILL YOU TAKE TWO NEGATIVE POINTS?'" [His wife:] "ARE YOU GOING TO END YOUR BOOK LIKE THAT?" [He:] 'NO…'" (466). By emphasizing that the book is all but finished, the protagonist voices a concern that also structures the graphic novel: change is not a once-and-for-all event but a process that includes setbacks and contradictions. As Squarzoni stated in a 2014 interview with the writer and critic Sarah Bernard,

> there are all different kinds of solutions that all mean something different politically. Certain solutions are dead ends, certain solutions drive a vision for society that I don't really agree with, and certain solutions drive a vision for society that I do agree with, but that still causes climate change. This is what I want to explore.[39]

Squarzoni uses the last pages of his book to sketch out one such solution via the voice of his alter ego, who is shown wandering through a snow-covered Ardèche, pondering what he has learned during his "journey through the science." Despite still heating with coal, this thinly disguised Squarzoni has undergone a deep and unfinished transformation: after learning about global warming and working on a graphic novel about this learning experience, he longs for an "INTELLECTUAL ARCHITECTURE" (443) to frame and structure future steps to bring environmental catastrophe to a halt.

His list of lessons learned includes that "MORALIZING ABOUT INDIVIDUAL BEHAVIOR," "EMBRACING GREEN CAPITALISM," and "BACKING THE IDEA OF SOLIDARISM" (443) by Léon Bourgeois[40] are not enough. Whether Bourgeois's political concept of solidarism is useful in combatting climate change goes beyond the aims of this chapter. What I nevertheless wish to draw attention to here is how strongly the ending insists on expert knowledge from fields other than the natural sciences, how explicitly it calls for politico-philosophical models and concepts. It is also remarkable that the text warns against depoliticization through academization. The narrator-protagonist wants to go beyond "NICE QUOTES" and also "BEYOND BACKING THE IDEA OF SOLIDARISM WITH ACTUAL POLITICAL AND LEGAL POWER" by "PROVID[ing] A FRAMEWORK FOR KEEPING CONCERN FOR SOCIAL JUSTICE AND ECOLOGICAL OBLIGATIONS IN MIND" (443). He calls for concrete measures to "REINVENT A CIVILIZATION" (440) and turns directly to the collective when it provocatively addresses the biggest issue of all: "CAN WE DO IT?" (443).

To facilitate a (positive) answer, the book turns to a group of economists[41] who work on "DEVELOPING SUSTAINABLE GLOBALIZATION AND

ECOLOGICAL ALTERNATIVES" (470). They all confirm that one "CAN" reduce energy to a massive extent, and demand and call for an economy that favors "PUBLIC SERVICES," "COLOSSAL INVESTMENT" in "CHANGING OUR TRANSPORTATION SYSTEMS, OUR HABITATS, OUR URBANIZATION" (445), to name just a few options. But they also insist that in order to be successful, transformation efforts depend on a collective change of mind, and way of life, in rich, industrial societies.

As this chapter has shown, these profound transformation processes take far longer than they should, and to make things even worse, they are steeped in ambivalence. Today, ambivalence is often considered unproductive, if not harmful, because it slows down the fight against global warming, which humanity is about to lose. By accepting ambivalence as a fact, *Climate Changed* manages to show, via a series of variations, how "hope" can thrive as long as human curiosity prevails. By challenging established perspectives and structures of feeling, it insists that cultural change is possible and a necessary, if slow, part of the solution. Despite its scientifically grounded negativity, the book turns out to be a surprisingly orientating and invigorating read that helps recognize and manage the ambivalences that threaten to paralyze initiatives against human-made climate change. By normalizing contradictory behaviors and mental ambiguities, by making room for moments of uncertainty, and by weaving the either-or-logic of scientific knowledge systems into an overall "chaotic" and ambiguous text, it helps to "defragment" the mind and find deliberately undefined "pockets of hope and possibility" where they are least expected. Furthermore, by playfully combining, or interrupting, the relationship between text and image, and between one sequence and another (with the help of unexpected "little stories"), *Climate Changed* privileges the gaps, contradictions, and simultaneities that are the unwanted "other" of modern, mass-mediated life. This makes for a surprisingly positive reading experience, as the book allows readers to occasionally "TAKE A BREAK" (96) from climate guilt and time pressure and cultivate a self-compassionate and positive approach to environmental transformation. Instead of dramatizing helplessness, it emphasizes agency and encourages readers to continue looking for alternative ways of making sense of the past, present, and future, even if it seems too late. Moreover, while it maintains that overarching political or technological concepts are indeed available, its own contribution is a cultural one that aims at the mental integration of undesired feelings (such as climate guilt) and a stronger focus on the small and big decisions we all make on a daily basis. By unveiling and challenging the norms, values, and structures that keep the societies of the Global North from responding adequately to global warming, *Climate Changed* moves beyond technical solutions (that it presents as valid approaches), radical renunciation, and sacrifice (that it identifies as insufficient and unrealistic but nonetheless an important exercise in personal transformation) and calls for (and seeks to inspire) a cognitive and emotional recoding on the levels of self, society, and decision-making political elites.

By sharing his personal, zig-zagging, and excessively slow and insufficient transformation toward CO_2 neutrality with the readers of *Climate Changed*, Squarzoni promotes a radical sense of hope and optimism that draws its strength from what's unknown, uncertain, and ambivalent. While the narrator-protagonist conveys that he, "PERSONALLY," could live with less but doesn't "BELIEVE" [that] "A RENEWED QUALITY OF LIFE DETACHED FROM CONSUMPTION, FROM THE RACE TO ACCUMULATE MORE [will] HAPPEN" (452), he readily admits that he "COULD BE WRONG" (457). This clarity about his own values, aligned with the dedication of the graphic novelist to promote long-term economic, political, and societal transformation through his own talent and expertise, shows that we can "begin" right here, at the end of this chapter, and against all odds.

Notes

1 Philippe Squarzoni, *Climate Changed: A Personal Journey into the Science* (New York: Abrams: 2014).

2 Comp. Karen Genschow, "Dissensuale Erzählungen," in *Closure*: *Kieler e-journal fuer Comicforschung* 1:11 (2014), 52, www.closure.uni-kiel.de/closure1/genschow.

3 The French version, *Saison Brune*, came out in 2012 (Paris: Edition Delcourt); the English version, on which this chapter relies, was published two years later.

4 For a theoretical discussion of how science authorizes its findings, see Marita Sturken and Lisa Cartwright, "Scientific Looking, Looking at Science" in *Practices of Looking: An Introduction to Visual Culture* (New York: Oxford University Press, 2001), 179–314.

5 Ann Gardiner, "Environmental Collapse in Comics: Reflections on Philippe Squarzoni's *Saison Brune*," in *The Discourse of Environmental Collapse: Imagining the End*, eds. Alison E. Vogelaar et al. (New York: Routledge, 2018), 65–87, 67.

6 Owain Jones, "The Ecology of Emotion, Memory, Self and Landscape," in *Emotional Geographies*, eds. Joyce Davidson, Liz Bondi and Mick Smith (Farnham: Ashgate, 2005), 191–204, 206.

7 Comp. Sturken, *Practices of Looking*, 1–44.

8 Concerning the difference between "showing" and "explaining," and the fruitful tension between the two in graphic novels, see Scott McCloud, *Comics richtig lesen* (Hamburg: Carlson, 1995), 146. English version *Understanding Comics: The Invisible Art* (New York: Kitchen Sink Press, 1993).

9 The terminology relates to Gérard Genette's theory of narration. *Die Erzählung*, 3rd edition (Paderborn: Fink UTB, 2010).

10 Alison E. Vogelaar, Brack W. Hale and Alexandra Peat, eds., *The Discourse of Environmental Collapse: Imagining the End* (New York: Routledge, 2018), 2.

11 Rebecca Solnit, *Hope in the Dark* (Edinburgh: Canongate, 2010) paraphrased in Vogelaar et al., 2.

12 Glenn Albrecht, "Solastalgia: a new concept in human health and identity," *Philosophy, Activism, Nature* (PAN) 3 (2005): 41–55.

13 Vogelaar et al., 2.

14 Zygmunt Baumann, *Retrotopia* (Cambridge: Cambridge University Press, 2017), 6–7.

15 Svetlana Boym, *The Future of Nostalgia* (New York: Basic Books, 2001), 50.

16 Boym, *The Future of Nostalgia*, 50.

17 Ibid.

18 J. David Bolter and A. Richard Grusin, *Remediation: Understanding New Media* (Cambridge: MIT Press, 2000), 45.

19 Walter Benjamin, *Illuminations*, trans. Harry Zahn (New York: Fontana, 1979), 260.

20 See McCloud, 78.

21 Title of the introduction to Zygmunt Bauman's *Modernity and Ambivalence* (Cambridge: Polity Press, 1991).
22 Baumann, *Modernity and Ambivalence*, 9.
23 Ibid., 6–7.
24 Ibid., 15.
25 Ibid., 9.
26 See Genschow, 52.
27 Remediations are refashionings of other media such as photography, advertisements, and film. See Bolter and Grusin, 273, and Genschow, 52.
28 I wish to point out that, from a feminist and BIPOC perspective, the causing of "climate fatigue" (disinterest in listening to the apocalyptic and complex elaborations that emerge from Western research institutions) also results from the book's failure to include international and BIPOC experts, whilst the marginality of female experts (there are only two of them, and they are both white) make this a very white, male text.
29 Regarding the manipulation of spatio-temporal expectations in graphic novels, see Harriet E. H. Earle, "Comics and the Chronotope: Time-Space Relationships in Traumatic Sequential Art," *HARTS and Minds*: *The Journal of Humanities and Arts* 1, no. 2 (2013): 2–13.
30 Jennifer Grubbs explains the emergence of the term "terrorism" in the context of environmental protest movements as a "rhetorical move" designed to exploit anxieties connected with terrorism discourse around 9/11." *Ecoliberation: Reimagining Resistance and the Green Scare* (Montreal: McGill-Queens University Press, 2021), 6.
31 Grubbs, *Ecoliberation*, 8.
32 The text uses quotation marks here to emphasize the cultural constructionist character of these terms that much of populist discourse depends upon.
33 "Climategate" refers to the 2009 political and media scandal linked to the hacking of thousands of emails from the University of East Anglia's Climate Research Unit. Climate change deniers decontextualized those emails to accuse leading scientists of faking evidence of human-caused global warming. See, e.g., Robin McKie, "Climategate 10 years on: what lessons have we learned?" *The Guardian*, November 9, 2019, www.theguardian.com/theobserver/2019/nov/09/climategate-10-years-on-what-lessons-have-we-learned.
34 Concerning the title, see " 'Saison brune': une BD sur le climat et notre 'schizophrénie'," *20 Minutes*, March 27, 2012, www.20minutes.fr/planete/905579-20120327-saison-brune-bd-climat-schizophrenie, and, also, Squarzoni, *Climate Changed*, 380.
35 A great introduction to basic design decisions in comics and graphic novels is the "layout" section in Chris Gavaler's blog "Analyzing Comics 101" at www.hoodedutilitarian.com/2015/12/analyzing-comics-101-layout/, accessed July 13, 2023.
36 Different meanings emerge depending on whether one reads the panels in columns or rows, and on whether one reads from left to right or from right to left. See Gavaler for an overview of common layouts and viewing paths through them. To learn more about the implied reading process, see Earle, "Comics and the Chronotope," 3.
37 Gavaler terms this a "column" reading path.
38 In 2012, when *Saison Brune* was published, there was still hope that reaching this margin could be prevented.
39 Sara Bernard, "This Graphic Novelist Tells the True Story of Climate Change," *Grist*, October 17, 2014, https://grist.org/climate-energy/this-graphic-novelist-tells-the-true-story-of-climate-change/.

40 The French politician and Nobel Prize winner Léon Bourgeois (1851–1925) was one of several scholars who have promoted, among other things, redistribution as a form of solidarity that recognizes the interdependencies that mark modern societies in a very material way. Hermann-Josef Große Kracht has published a very informative review (unfortunately in German only) on Bourgeois' popularity in France and the 2020 German translations of some of his texts: www.soziopolis.de/ueberfaellig-aber-missglueckt.html.

41 The featured experts are Geneviève Azam, Jean-Marie Harribey, and René Passet.

Concluding Remarks and Survey of the Contributions

Hubert Zapf

The call for the ecological transformation of society is one of the more commonly shared leitmotivs of current environmental debates. Based on scientifically corroborated knowledge about anthropogenic planetary disruptions that threaten human and nonhuman ecosystems, it demands an immediate, consequential, and sustainable response by individuals and societies. In this sense, an "ecological imperative" (Hans Jonas)[1] has superseded the Kantian "categorical imperative" in contemporary environmental thought. It is from this ostensibly self-evident ecological imperative, involving a "principle of responsibility" ("Das Prinzip Verantwortung") for the long-term survival and well-being of life on earth in the face of socio-environmental degradation and injustice, that current forms of environmental activism, as well as academic research fields such as the Environmental Humanities, have taken their motivation and inspiration.[2]

However, while the diagnosis of the current environmental crisis is based on widely accepted factual knowledge, its interpretation and the question regarding its possible mitigation or redress are much more complex and fraught with ambivalences. The International Doctoral College *Re-Thinking Environment: The Environmental Humanities and the Ecological Transformation of Society*, jointly organized by the University of Augsburg and the LMU Munich, within whose framework the present volume originated, takes both of these aspects seriously: the urgent necessity of socio-ecological transformation and the complexities and ambiguities emerging from within its research agendas themselves, as well as in the translation of environmental knowledge into environmental practice and policies. The volume focuses on this field of tensions within and beyond environmental research from different thematic and disciplinary angles and thus reflects the multifaceted and multidisciplinary potential of the Environmental Humanities in addressing the perspectives and challenges of socio-environmental change.

An obvious critical objection to such attention on ambivalence and complexity might be that it contributes itself to resisting the speedy socio-ecological transformation that the Environmental Humanities purport to enable. Nevertheless, this is a paradox that characterizes all serious scholarly research, and even if it addresses urgent societal issues, it is not itself an unmediated form of sociopolitical engagement such as, in this case, environmental activism. It rather helps to explore the

DOI: 10.4324/9781032627984-17

inherent diversity and potentially conflictual interrelation of the various factors contributing to the environmental crisis and the most promising means of coping with it. In this way, it also supplies actors involved in socio-ecological transformation in sociopolitical practice, as well as the wider public, with a more differentiated, multi-perspectival—and thus, in fact, a more realistic and effective—basis for their active engagement.

If this makes environmental change more demanding, the call for immediate transformative action, as plausible as it may be in principle, risks becoming reductionist when it acts upon moral convictions rather than the awareness of multiple entanglements and inevitable tensions as basic features of socio-ecological processes. On the one hand, manifold environmental projects on the local, regional, national, and global levels are carried out by engaged citizens, NGOs, political organizations, and a diverse range of responsible agents in the public arena that are increasingly aware of such tensions between different agendas and, indeed, of "inner-ecological conflicts,"[3] e.g., between renewable energy technology and nature protection, CO_2 reduction and biodiversity loss, or between conservationist and decolonial agenda, and of the necessity to take into account and cooperate across such conflict lines. On the other hand, even highly impactful forms of environmental activism with a worldwide appeal, such as Greta Thunberg's Fridays for Future initiative, which has motivated many, especially young people to join the environmental cause and has put pressure on international politics, has also unwittingly found itself in the middle of such ambivalences and inner-ecological conflicts. When Thunberg advocated nuclear power to help reduce CO_2 emissions, for example, her followers (especially in Germany) were rather disenchanted, since in Germany the DNA of the Green movement, more perhaps than in other parts of the world, has been defined from its origins in the 1970s by its fight against the incalculable risks of nuclear energy.

Critical awareness of ambivalence and complexity, which is foregrounded in the contributions of the present volume, thus implies no avoidance of but an inherent requirement for transformative ecological thought in adequately responding to the urgency of current environmental problems and in developing sustainable alternatives for the future. The Environmental Humanities are a pluri-versal and multi-perspectival epistemic field, which recognizes the limitations of only one view or source of evidence and therefore requires transdisciplinary collaboration in an open, constantly self-revising process of focused exploration. In this process, nevertheless, valuable insights are gained that are of considerable significance for the ways in which ecological thought and ethics can be more effectively related to ecological practice, and vice versa.

What the contributions to the volume also show is that ambivalence as a concept is itself ambivalent.[4] In a negative sense, it designates a lack of referential transparency and ethical consistency, including internal contradictions and conflicts of interests within purportedly "clean" environmental projects such as in the case of economic greenwashing or in neocolonial forms of resource exploitation in the name of development and progress. In this sense, it is a *critical-deconstructive concept* for bringing these contradictions to light, which means uncovering the wider, often unacknowledged psychological and sociopolitical structures in which such activities are implicated.

In a positive sense, ambivalence suggests awareness of relational tensions as intrinsic features of an environmental epistemology and practice, in which no single issue or position can be viewed in isolation from others. As such, it is a *generative epistemic concept* in Environmental Humanities research that designates a non-binary thinking of connectivity-within-diversity, an inclusionary *both-and* rather than an exclusionary *either-or* form of thinking. These two types of ambivalence often interact, with shifting preference, in the contributions of this volume, and thus they constitute not just an ambivalent but a pluri-valent approach to its overarching topic.

The volume demonstrates the explorative value of this approach in two main parts, one part focusing on the conceptual level, the other on the practical level, with mutual resonances between the two of them throughout the contributions. The opening chapter of Part I, *Conceptual Facets*, is Matthias Schmidt's chapter "Ecological Transformation and the Implications of Spatial Scale." From the perspective of human geography and political ecology, it is an example of the use of ambivalence as a tool of critical analysis. It addresses the widely observable contradiction between a commonly shared "articulated consent" to ecological transformation and a simultaneous, often concealed resistance to such transformation "in large sections of society, politics, economics and science." Schmidt suggests "spatial distance" as one important factor that ties together a critical view of globalization, the limited horizon of human perception, the lack of adequate institutions, and the abstract scale of global climate change into a latent lethargy mode preventing an active response to climate issues or empathy for problems in far-distant regions.

Dietmar Süß addresses ambivalences of eco-political engagement from a historical perspective in "Climate Solidarity, Time, and Ambivalence: On a New Term and Its Historical Legacy." The author traces the widespread contemporary concept of climate solidarity in its (dis-)continuities with the tradition of social solidarity, focusing on developments in Germany from the 19th century to the present. In a detailed reconstruction of the conflictual history of solidarity in labor movements, the Social Democratic Party, and more recent Green movements in Germany, Süß points out how ambivalences in earlier historical constellations, such as "between solidarity as a *political concept of struggle* and solidarity as a *mode of social integration*" (my emphasis), continue to shape contemporary debates on climate solidarity. This is also true of the "contentious relationship between social-democratic thought and modern environmentalism," which indicates conflicting visions of progress and growth in a tension between socio-economic and environmental priorities that need to be negotiated in the discourse of climate solidarity.

Marcus Wagner, in his chapter "Interactions of Efficiency, Consistency and Sufficiency across Levels: Innovation Tensions through the Lens of Paradox Theory," brings the perspective of Economics into conversation with the Environmental Humanities in a similar combination of a critical and a generative use of ambivalence. With reference to paradox theory, he addresses the paradoxes and ambiguities arising from the observation that "corporate sustainability actions can still be complicit in creating global environmental damage." Revisiting impact-population-affluence-technology (IPAT) to calculate the human impact on the environment, Wagner argues for complementing "efficiency" as a technological factor with "sufficiency," i.e., "individuals' willingness to abstain or abdicate from consumption," and "consistency," i.e., "the avoidance of resource overuse, high levels of circularity, and the usage of substances that already exist in nature." That way, humanist and ecological values are integrated into the techno-economic equation, thereby making tensions and ambivalences explicit which are otherwise ignored or suppressed at the price of self-deception or the ideological coverup of hidden power interests. Inspired by the "ecological imperative" of transgenerational justice (Jonas), Wagner proposes a multi-scalar, *both-and* mode of environmental economic thinking which reflexively incorporates the paradoxical tensions between techno-economic and humanist perspectives on various levels of agency. He does so in search of more complex—and thus more viable—paths of sustainable transition, fully aware that such transition always involves "gains and losses which challenge historically developed and reproduced power distributions and infrastructures."

From a decolonial literary ecology, Victor A. Ferretti, in "Transformation and/of 'Colonial Tropes.' Latin American Narrative Palimpsests," critically analyzes the ambivalences and hidden agendas of socio-environmental conflicts in Latin America as a systemically marginalized area in the Global South, due to "systemic one-sidedness" and neocolonial "exploitation dynamics." Latin America in this view is a palimpsest of European "exploitation marks" that continue to shape apparently environmentally friendly projects such as e-mobility. Ferretti explores the literary response to these ambivalences from a historical distance in the example of *novelas de la selva* ("jungle novels"), which expose the detrimental effects of the transnational rubber industry in the 19th and 20th centuries on local people and the environment. The ecocritical approach he applies to these literary substance narratives aims at "avoiding the notion that required socio-ecological transformation turns unintentionally into an 'imperialistic' trope"—thus emphasizing the importance of ambivalence as a deconstructive epistemic mode in addressing socio-ecological transformation.

In the final chapter in this first part, namely, "Toxic Commons and the Politics of Ambivalence: Re-imagining Toxic Legacy Sites," Simone Müller and Angeliki Balayannis place thinking and living with ambivalence at the center of their contribution on how to deal with toxic legacy sites. Combining the perspectives of environmental history and human geography, they propose "Toxic Commons" as a new concept in the Environmental Humanities. The idea enables scholars and practitioners to "engage with the ambivalences of toxic exposures that equally divide and bind humanity across the planet." The authors frame their chapter with reflections on the deeply ambivalent site of the Rocky Mountain Arsenal National Wildlife Refuge near the city of Denver, whose mission was to restore an original American natural environment but which as a former Army site turned out to be heavily contaminated by chemical and other toxic waste beyond its human-imposed boundaries. In the face of this unruly, transboundary nature of toxic waste, the authors argue for a "relational epistemology" that overcomes "binary conceptions of containment" in actively embracing a "politics of ambivalence." The concept of Toxic Commons takes the inseparability and planetary interconnection of toxic pollution as a reality and considers toxic legacy sites as a "shared heritage of humankind," stretching across the boundaries of nations and geographies. This kind of ambivalence combines a negative element—the deconstruction of illusionary purity—with a positive element—the affirmation of responsible relationality. In this way, it aims not only at a more realistic assessment of omnipresent toxicity, but also at establishing an ethics of care for human and nonhuman lives from an anti-capitalist vision of both communal and planetary awareness and cooperation in addressing toxicity, power asymmetries, and environmental justice.

The eight contributions in Part II, *Ambivalences in Practice*, illuminate the overarching topic of ambivalence in concrete examples from environmental practice in various material, discursive, medial, and regionally specific contexts. As stated before, they also resonate with the chapters in Part I in the (self-)critical exploration of the tensions between discursive-conceptual claims and the practical challenges of socio-ecological transformation.

The opening chapter of Part II by Sebastian Lundsteen, "In Praise of Ambivalence: Reflections on Experiences of Pollution and Remediation," explicitly ascribes ambivalence a positive meaning in its role in negotiating the complexities of socio-ecological transformation. Lundsteen takes as his example the remediation project of a submerged chemical waste depot in western Denmark, Hofde 42-depotet, initiated by the Danish government to contain the dangerous toxic waste deposited there by a chemical company over many decades. However, the unstable site of the deposit near a stormy coastline has raised doubts and anxieties among activists and concerned citizens about the viability of the technological remediation project. Starting from this observation, the chapter builds its reflections on a commendation of ambivalence as a "tool for analyzing human experiences of environmental transformation." As Lundsteen points out, ambivalence in this sense

> signifies neither undivided optimism nor pessimism or even indifference towards a given process, but it holds both positions by expressing reluctance towards certainty and facts while still acknowledging them … . [A]rticulations of ambivalence challenge myopic tech-optimistic solutions as dominant epistemic regimes. As such, ambivalence can reveal otherwise hidden power structures and specific historical trajectories.

Iris Boroway, in "Global Waste: On the Ambivalence of Wealth, Health, and Contradictory Development Models," describes the intrinsic ambivalences of the globally prevalent economic development model that has "both improved and harmed human lives" through its mode of mass production and consumption, on the one hand, and its generation of large-scale quantities of waste, on the other. It has been undeniably successful, as the author illustrates in the development of China in the past few decades into a leading economic power with spectacularly improved rates of wealth, health, and standards of living, albeit it has also produced an unprecedented quantity of waste, the disposal of which has become increasingly intractable. The chapter discusses two influential, alternative models that have sought to respond to this challenge. The first is the *circular economy*, which aims at imitating in the economic sphere natural material recycling processes, and thus at avoiding or minimizing waste within the existing system. The second model, *degrowth*, defines human well-being in cultural-ethical rather than expansive economic terms and thus tries to separate waste from the mass production/consumption system altogether. Nonetheless, these two models produce their own ambivalences. While the circular economy is more easily compatible with the current economic system, it tends to accommodate itself to, rather than truly alter, the problem of large-scale waste generation. And while degrowth means a more radical step toward a different socio-economic system, its very radicalism seems to weaken more general public support. Yet, the author sees this double ambivalence not only negatively, but also rather as a chance to test out new "overlapping practices, experiments and developmental pathways, which may offer a good chance in the future of breaking out of an unhealthy connection between the good life and waste."

In "Farming the Wind: Aeolian Politics and the Sacred Desertscapes (Orans) of Rajasthan," Nisha Paliwal and Dibyadyuti Roy take the example of large-scale wind farming projects encroaching on the ecosystems of indigenous communities in the state of Rajasthan in India to expose the ambivalences of what they call "aeolian politics." As the authors argue, the normative agenda of global CO_2 reduction, codified in the UN Sustainable Development Goals and adopted by the Indian government, clashes with local realities overlooked in dominant narratives of green transformation. "The dreams of aeolian futures are [often] paradoxical, what are heavenly images for some are nightmares for others." The desertscape of *Aalaji ka Oran*, on which the authors focus in their case study, was declared in official government documents as a "wasteland" and thus a non-place that could be used for renewable energy industrialization, while in reality it was a place of well-adapted inhabitation and sustainable human-environment interaction, in which the cattle economy, water resource management, and biodiversity protection were part of long-standing cultural traditions. Places like this, called *Orans*, were sacred sites at which the veneration of local deities and of nature was encoded in rituals and folk tales that continually preserved an intimate connection between human and nonhuman ecosystems. Paliwal and Roy show how not only the livelihoods, but also the environmental knowledge of indigenous communities encoded in such traditions are threatened by the side-effects of large-scale wind energy infrastructure infiltrating their areas. The authors, however, end their chapter on a positive note. The organized resistance of local activists against the pressures of this kind of "aeolian politics" just recently led to a partial mitigation of problems when *Aalaji kan Oran* was finally removed from the "wasteland" category in governmental revenue records. In its combination of historical surveys, the close reading of documents, and humanistic fieldwork, the chapter thus shows how "the reflective analytical frameworks championed in the humanities can lead to contemplative research and co-produced action on key environmental challenges."

Also, with a view to the impact of global modernity on local communities in the Global South, Sule Emmanuel Egya's chapter "Global South Ambivalences of Transformation? Literature, Extractive Capitalism, and Literary Militancy in West Africa" exposes discrepancies between modern ideologies of progress and exploitative extraction practices in West Africa, which lead to disastrous consequences for local living conditions through oil spills, toxic pollution, sickness, and the loss of vital interrelations between human and nonhuman ecosystems. These socio-ecologically destructive practices of "petrodollar capitalism," as Egya argues, provoke powerful voices of resistance in contemporary African poetry and narrative prose. Intellectuals and writers on the continent increasingly revolt against this doublespeak of governments and transnational corporations and adopt forms of "literary militancy" in a strategy of indigenous self-empowerment, as Egya exemplifies in Nigerian writer G'Ebinyo Ogbowei's poem "Marsh Boy" and Cameroon-American Imbolo Mbue's novel *How Beautiful We Were*. Ambivalence is not a solution but a problem in these forms of ecological protest literature; it is perceived instead as an ideological tool and an obstacle to real socioecological transformation.

The subsequent chapter by Rafael Hernandez Westpfahl, "Mining for a Low-Carbon Economy? Articulations by the Mexican Corporate Sector," likewise exemplifies a critical use of ambivalence by analyzing the ways in which corporate industry—in this case in Mexico—tries to conceal existing contradictions in the ecological transformation promised by renewable energy technologies, especially focusing on "green" minerals. On the basis of public documents released by the mining companies, Westpfahl demonstrates how their discourse strategies aim at obscuring the environmental and social costs of mineral production in favor of unambiguous narratives of sustainable development and decarbonization. This rhetoric of green transformation, as the author argues, serves rather to distract from the hidden agenda of these companies to continue their profit-oriented operations within the new label of decarbonization while in fact still heavily relying on fossil energy.

Decarbonization ambivalences are also the subject of Rocío Hiraldo's chapter "Integrating Labor, Environment, and Climate? (Dis-)connections in the Spanish and Portuguese Energy Decarbonization." Like Dietmar Süß in his chapter on climate solidarity, Hiraldo focuses on tensions between social and environmental priorities as traditionally represented by trade unions vs environmental organizations. Moreover, at the same time, he shows how this conflictual pattern is becoming less entrenched in concrete cases of transforming to decarbonization in Spain and Portugal. He also investigates the responses to coal plant and oil refinery closures, as well as to large-scale renewable energy projects in the two countries, by workers and labor organizations, on the one hand, and by environmental activists and organizations, on the other. In evaluating primary and secondary data collections, he arrives at the conclusion that there is a growing tendency on the labor side to include environmental concerns, while environmentalists include socio-economic concerns in their agendas. Hidalgo thus diagnoses an increasing convergence of environmental, labor, and climate goals in the decarbonization process in these two countries, which could be a model for other parts of the world as well.

In their chapter "Bioplastics versus Conventional Plastics: An Analysis from a Sociological, Ethical, and Educational Perspective," Ulrike Ohl, Maria Backhouse, Kerstin Schlögl-Flierl, Anja Kalch, and Helena Bilandzic combine a critical-analytical with a generative concept of ambivalence in a collaboration between the three disciplines of sociology, environmental ethics, and didactics of geography. From a sociological and an ethical perspective, they examine "ecological trade-offs" and "ambivalences of bioplastic" as an apparently ecofriendly but in fact problematic alternative to plastic waste. However, they do not try to eliminate this ambivalence when they subsequently adopt the perspective of environmental didactics and instead consider it as motivational energy for democratic education which, rather than imposing easy answers, helps empower students "to make their own decisions." The chapter thus explicitly makes the ambivalence between the "urgency" of environmental transformation and the aim of "transparent, participatory knowledge transfer" a key point in its multidisciplinary project of environmental research and education.

In the concluding chapter, "Against All Odds: Managing Ambivalence in Philippe Squarzoni's Graphic Novel *Climate Changed*," Kirsten Twelbeck explores this productive potential of ambivalence in a literary example, namely, the graphic novel *Climate Changed* (2014) by Philippe Squarzoni. Interweaving a wide spectrum of scientific climate communication and personal responses to climate change, the novel, as Twelbeck argues, actively advocates emotional and behavioral ambivalence rather than provides clear-cut answers or technological solutions as a way of dealing with the environmental crisis. In its genre-specific interplay between text and image, the graphic novel creates a counternarrative to conventional development narratives in the form of an "anti-quest" that foregrounds "ambivalence, incompatibility, and tension" and, avoiding unrealistic happy endings, nevertheless creates paradoxical spaces of possibility for hope "against all odds," as the chapter's title suggests.

All in all, the volume convincingly demonstrates the importance for the Environmental Humanities of dealing with ambivalence and complexity as inevitable features of socio-ecological transformation. In contributions from various disciplines across the range of the humanities and social sciences, it offers a rich source of insights into the pluri-versal field of concepts, areas, practices, and perspectives of socio-ecological transformation, assessing both their promises and potentials and their internal tensions and ambiguities in the interest of a more viable, more self-aware, and more democratic-participatory transition to a sustainable society and culture. To acknowledge ambivalence as an explorative concept within transformative ecological thought means complexifying apparently straightforward answers and discursive constructions of binaries. It enables the critical exposure of socio-ecological injustice, inequality, and power asymmetries obscured in conventional narratives of progress and development. At the same time, it advocates a relational epistemology and ethics of human-environment interdependence as a fundamental new orientation in academia as well as in the larger society and culture.

Thinking with ambivalence is thus not a weakness but a strength of the humanities and social sciences. In the contributions to this volume, political, social, cultural, literary, ethical, economic, and educational dimensions of responding to the environmental crisis are foregrounded as a necessary counterpole to merely managerial or techno-scientific solutions, in a multi-perspectival approach that combines a critical with a productive contribution to responsible environmental knowledge and practice.

Notes

1 Hans Jonas, *Das Prinzip Verantwortung: Versuch einer Ethik für die technologische Zivilisation* (Frankfurt: Suhrkamp, 1979), trans. *The Imperative of Responsibility: In Search of an Ethics for the Technological Age* (Chicago: University of Chicago Press, 1984).

2 See, for example, the Bern Sinergia Project, "Mediating the Ecological Imperative," accessed July 9, 2024, www.ecological-imperative.ch/.

3 Jens Soentgen, "Am Ende des Zwei-Grad-Ziels: Für ein neues Denken im Klimadiskurs" ["At the End of the Two-Degree Goal: For a New Thinking in the Climate Discourse"] *Merkur* 74 (2020): 22–33.

4 Jens Soentgen, "Am Ende des Zwei-Grad-Ziels," 33: "A new thinking in climate discourse involves more awareness of the ambivalences of our solutions. Ambivalence [...] means that something has simultaneously positive and negative aspects. Such ambivalences are inconvenient, one tends to ignore or belittle them. It is however important to accept them and find ways of productively dealing with them" (my translation).

Index

Note: Figures are indicated by *italics*. Tables are indicated by **bold**. Endnotes are indicated by the page number followed by 'n' and the endnote number e.g., 20n1 refers to endnote 1 on page 20.

MIX
Papier | Fördert
gute Waldnutzung
FSC® C083411

Kolibri 360 GmbH
Ferdinand-Jühlke-Straße 7
99095 Erfurt, Deutschland
produktsicherheit@kolibri360.de